Estimation in Surveys
with Nonresponse

WILEY SERIES IN SURVEY METHODOLOGY

Established in part by WALTER A. SHEWHART AND SAMUEL S. WILKS

Editors: *Robert M. Groves, Graham Kalton, J. N. K. Rao, Norbert Schwarz, Christopher Skinner*

A complete list of the titles in this series appears at the end of this volume.

Estimation in Surveys with Nonresponse

CARL-ERIK SÄRNDAL
Université de Montréal

SIXTEN LUNDSTRÖM
Statistics Sweden

John Wiley & Sons, Ltd

Other Wiley Editorial Offices

John Wiley & Sons Inc., 111 River Street, Hoboken, NJ 07030, USA

Jossey-Bass, 989 Market Street, San Francisco, CA 94103-1741, USA

Wiley-VCH Verlag GmbH, Boschstr. 12, D-69469 Weinheim, Germany

John Wiley & Sons Australia Ltd, 42 McDougall Street, Milton, Queensland 4064, Australia

John Wiley & Sons (Asia) Pte Ltd, 2 Clementi Loop #02-01, Jin Xing Distripark, Singapore 129809

John Wiley & Sons Canada Ltd, 22 Worcester Road, Etobicoke, Ontario, Canada M9W 1L1

Wiley also publishes its books in a variety of electronic formats. Some content that appears
in print may not be available in electronic books.

Library of Congress Cataloging-in-Publication Data:

Särndal, Carl-Erik, 1937–
    Estimation in surveys with nonresponse / Carl-Erik Särndal, Sixten
Lundström.
        p. cm.
    Includes bibliographical references and index.
    ISBN-13 978-0-470-01133-1 (acid-free paper)
    ISBN-10 0-470-01133-5 (acid-free paper)
    1. Estimation theory. 2. Sampling (Statistics) 3. Nonresponse
(Statistics) I. Lundström, Sixten. II. Title.
    QA276.8.S2575 2005
    519.5′44 – dc22
                                            2005001255

British Library Cataloguing in Publication Data

A catalogue record for this book is available from the British Library

ISBN-13 978-0-470-01133-1 (HB)
ISBN-10 0-470-01133-5 (HB)

Typeset in 10/12pt Times by Laserwords Private Limited, Chennai, India
Printed and bound in Great Britain by TJ International, Padstow, Cornwall
This book is printed on acid-free paper responsibly manufactured from sustainable forestry
in which at least two trees are planted for each one used for paper production.

Contents

Preface

The preface of the 1963 edition of William G. Cochran's *Sampling Techniques* begins: 'As did the previous editions, this textbook presents a comprehensive account of sampling theory as it has been developed for use in sample surveys.' The reader, theoretician or practitioner, who hopes to find some theory and methods for the treatment of nonresponse must patiently await the concluding Chapter 13 of the book, where a few suggestions are given.

Sampling Techniques and other classical sampling texts bear witness to the important scientific advances made in survey theory during the first half of the twentieth century, although without perceiving a need to dwell extensively on estimation in the presence of nonresponse.

Today, nonresponse is a normal (but undesirable) feature of the survey undertaking. Accordingly, the underlying theory should recognize nonresponse from the beginning. Nonresponse can no longer be handled by an adjunct to a theory that would work well if it were not for certain less desirable realities. The present book attempts to build on these premises.

This change in perspective does not entail any criticism of earlier work. It simply reflects the changing times and the changing survey climate.

Writing more than 40 years ago, Professor Cochran had good reason for relegating nonresponse to the end of his book. The book is remarkable, not least because it served as a valuable guide to a generation of survey statisticians. It was about the theory of sampling, about controlling the sampling error, about estimators that work well under controlled randomized selection. Nonsampling errors are summarily mentioned in Chapter 13. They include errors due to nonresponse and missing observations, errors of measurement, errors in editing and tabulation. At that time, nonresponse was low and its effects negligible.

Even today, sampling error is often contrasted with nonsampling error. But it is a dichotomy that feels much less appropriate now than it may have been 40 years ago. For one thing, the distinction is lopsided: it places together in one category the errors attributable to all causes other than sampling and observing only a fraction of the finite population. Some of the nonsampling errors may indeed be greater than the sampling error.

To us, and no doubt to many others, it is unsatisfactory to have a theory that focuses on sampling error and handles nonsampling errors through 'adjustments' that one somehow appends to a theory meant for ideal conditions not met in practice. Today's reality includes high nonresponse and often imperfect frames.

In writing this book, we found inspiration in the extensive literature on non-response. Although further development is bound to follow, we believe that our book has something to offer to all categories of professionals engaged in surveys and statistics production, survey managers, subject-matter specialists, as well as specialists in statistical survey methodology.

The background knowledge required to assimilate the contents of this book varies between the chapters. Roughly the first one-third of the book is easy to follow for all categories; the rest requires more of a technical preparation.

Chapters 2 and 3 give a wholly nontechnical overview. They can be read with only a minimal background in statistical science. They emphasize general issues in the treatment of nonresponse and provide a general orientation in the field. Technical arguments and formulae are essentially absent.

Chapters 4–14 form a logical sequence in which one chapter builds on the preceding ones. The theory of estimation in the presence of nonresponse is developed from Chapter 6 onwards. Many of the notions of the classical survey theory are preserved. There is a finite population of identifiable elements with which one can associate individual probabilities of being selected and of responding. The classical randomization theory is obtained as a special case, namely, when the nonresponse is reduced to nil.

The contents of Chapters 4–14 can be described as moderately technical. To fully appreciate these chapters, readers should have been exposed to at least one thorough course in statistical inference, including principles of point estimation and confidence intervals. Familiarity with linear models and regression analysis, including some linear algebra, is important background. It is a considerable advantage also to have followed one or more courses in modern survey sampling theory and techniques.

Calibration is the unifying concept that keeps the chapters together. Calibration is a technique for computing weights to be used in estimation, given an input of auxiliary information.

Persons working actively with survey methodology, in statistical agencies or survey institutes, should be able to easily follow the whole book.

Readers with practical experience with surveys and statistics production are in a favoured position. They can identify with many issues raised in the book. Their practical background will facilitate their reading.

A reader who is a first-year graduate student at university may start from a background derived mainly from a set of undergraduate statistics courses and with little or no exposure to the reality of surveys. Nevertheless, the book should find use in a graduate course within a university statistics curriculum.

Some more technical material – notably derivations and lengthier proofs of certain propositions – is placed in end-of-chapter appendices. These need not be read or understood if the primary objective is to see the practical utility of the methods.

This book evolved from a Current Best Methods (CBM) manual which we wrote in 1999–2001 for Statistics Sweden. Its title is *Estimation in the Presence of Nonresponse and Frame Imperfections*. A far-reaching revision and extension of the contents of that manual was necessary to meet the objective that we and the reviewers set for this book.

The contents have a definite practical orientation. On the other hand, it would not be appropriate to call the book a handbook. A practitioner should perhaps not hope to find, on a certain page of the book, a tailor-made answer to a particular problem encountered in a survey. He or she should always be prepared to adapt the techniques in this book to fit the context of his or her own survey. This may require familiarity with probabilistic evaluations of expected values and variances, with techniques in regression analysis and with other knowledge mastered by survey methodologists.

We gratefully acknowledge the support of Statistics Sweden in the preparation of this text. Three anonymous reviewers engaged by John Wiley & Sons, Ltd commented on the proposed content. We are indebted to them for their valuable comments, which provided a clear direction for our work.

Ottawa   Carl-Erik Särndal[*]
Örebro   Sixten Lundström[†]
December 2004

---

[*]**Carl-Erik Särndal**, Ph.D., is Consultant to National Statistical Agencies and was formerly Professor, Université de Montréal
[†]**Sixten Lundström**, Ph.D., is Senior Methodological Adviser, Department of Population and Welfare Statistics, Statistics Sweden

CHAPTER 1

Introduction

Nonresponse has long been a matter of concern in statistical surveys. The attention paid in the literature to nonresponse and its effects has increased dramatically in the last few decades. This is in part the effect of a hardened survey climate and a decreased willingness on the part of the general public to cooperate and deliver the data solicited by statistical agencies or by private survey institutes.

Nonresponse is a normal but undesirable feature of a survey. There is complete agreement that nonresponse can severely harm the quality of the statistics computed and published in a survey. Both producers and users are well aware of this. Everyone, regardless of his or her training in statistical science, can understand why nonresponse can harm survey results. For example, if women show a considerably greater inclination to participate and to respond to a survey than men, then a random sample of persons will have an overrepresentation of females among the respondents. Downwardly biased estimates will result for the whole population total of any variable, such as income, on which women score lower on average than men, *unless* the estimation techniques can rectify the tendency to bias. The spotlight thus turns on to the estimation techniques that manage to perform well despite some (or even considerable) nonresponse.

There is a copious literature on nonresponse, much of it dating from the past two decades. This literature examines two different (but complementary) aspects: the prevention or avoidance of nonresponse before it occurs; and the estimation techniques needed to properly account for the nonresponse that has nonetheless occurred. The literature often refers to the latter activity as nonresponse adjustment. In preparing this book, we have benefited from important parts of this literature.

If nonresponse could be easily prevented, or virtually eliminated, the issue would disappear. Considerable resources are therefore spent on improving data collection procedures, so that, if not totally eliminated, nonresponse will at least be substantially reduced. Follow-up of nonrespondents tends to be costly. Statistical agencies are well aware of this; some have formulated their own guidelines for cost-effective data collection.

The methods for nonresponse prevention draw on a body of knowledge from within the behavioural sciences. This is natural because the data collection involves contact with respondents. It is necessary to overcome scepticism and to promote a positive attitude to the survey. Motivational factors play an important role. Innovative research on survey motivation is reported in recent articles and books. That body of literature lies outside the scope of this book.

This book concentrates on the second question, the estimation techniques suitable for survey data affected primarily by nonresponse, and secondarily by frame coverage problems. Once data collection is concluded, those in charge of a survey are often forced to recognize that a considerable nonresponse has occurred. Some desired data are missing. Survey managers face deadlines and must stay within an allotted survey budget. The data collection phase must come to a halt at a definite point in time. Estimation, inference and the publication of results must go ahead, based on the data gathered up until that moment in time.

Estimation and inference rely on a body of knowledge from statistical estimation theory. This holds also for data affected by nonresponse. The reasoning relies on mathematical and statistical notions.

In a perfect world, a survey has no nonresponse. All selected elements cooperate and deliver all of the requested data, with no measurement error. In that perfect but non-existent world, a survey has only sampling error.

Sampling gained popular acceptance in the 1930s and 1940s as a much less costly alternative to total enumeration of the finite population. It was a major scientific achievement to propagate the idea that sampling error could be effectively controlled through efficient sampling design. Nonresponse rates were low during the early part of the twentieth century. But as time went on, nonsampling errors became increasingly recognized as trouble spots for sample surveys. Nonresponse error came to be the most publicized of the nonsampling errors. Nonresponse adjustment techniques were developed. The recognition of other nonsampling errors, such as frame errors and measurement errors, was another significant development.

The challenge facing today's survey statistician is to make the best possible estimates based on (i) the data at hand when the data collection phase must necessarily stop and (ii) any relevant auxiliary information that can help the estimation, information about the population and its elements, whether they be respondents or nonrespondents. This book presents theory and methods of estimation under these conditions. We seldom use the term 'nonresponse adjustment'. There is no definite need for it in the approach that we take. Besides, the term has a certain negative ring. It suggests that well-conceived techniques have somehow failed to meet the challenge, so that they need to be repaired or patched up. We prefer to view the task as one of estimation in a broader context, namely, the best possible estimation in the presence of nonresponse and coverage problems.

The techniques discussed in Chapters 6–14 are held together by the unifying concept of calibration. Calibration estimation is a weighting method; the weights are computed with an input of auxiliary information, as explained and illustrated in Chapters 6–8. The weights can effectively compensate for the nonresponse, provided that powerful auxiliary information is identified.

Simplicity and transparency are attractive features of the calibration approach. The technique is easy to apply. Based on a stated input of auxiliary information, the weights are computable by existing software.

A central question arises: How do we build an auxiliary vector powerful enough to reduce the nonresponse bias of the estimates to minimal levels? How do we identify the auxiliary variables that are most likely to contribute significantly to bias reduction, so that they should be put into the auxiliary vector? We examine these questions in considerable detail, particularly in Chapters 9 and 10.

Why, then, is it so important to actively and energetically seek reduction of nonresponse bias to 'minimal levels'? One answer is that the statistical inferences made in a survey – by confidence intervals, for example – run the risk of being invalid. As Chapter 11 notes, this ever present risk conveys 'a message of serious consequence for all statistical inference from data affected by nonresponse'.

Bias in survey estimates can go undiscovered for long periods of time. Examples from important regularly repeated surveys have shown how the level of the estimates has changed drastically when new important auxiliary variables have been brought to bear on the calibrated weighting. The change in level is attributable to a significant reduction of the nonresponse bias that probably affected the estimates in the past. Examples are mentioned in Section 3.2.

Imputation is recognized as the other traditional tool, in addition to weighting, for treatment of nonresponse. Chapters 12 and 13 discuss how imputation finds use in combination with weighting, derived through calibration.

Chapter 14 deals with calibration techniques for surveys with frame deficiencies.

An important aspect which we leave unexplored is resource allocation in a sample survey. A managerial question, in a statistical agency and in other survey institutes, is how to split limited available resources between efforts to prevent nonresponse from occurring and efforts to obtain high-quality estimates despite a nonresponse that has already occurred. The literature to date does not offer conclusive theory and advice on this split.

We do not recommend a use of *ad hoc* formulations of an 'admissible extent' of nonresponse in a survey. Rules or advice are sometimes proposed, stating that a minimal response rate, say 80%, must be ascertained in a survey. Not only may the adherence to such rules prove costly from a data collection point of view, it can also result in a far from optimal use of the limited resources. The harmful effects of nonresponse, notably the bias in the estimates, can vary considerably from one survey to another. Adherence to fixed rules appears to be futile. An important question instead is to evaluate whether or not the survey results are likely to be severely distorted by the nonresponse. If they appear to be so, appropriate action is required.

CHAPTER 2

The Survey and Its Imperfections

2.1. THE SURVEY OBJECTIVE

The objective of a survey is to provide information about unknown characteristics, called parameters, of a finite collection of elements, called a population,
such as a population of individuals, of households, or of enterprises. A typical
survey involves many study variables and produces estimates of different types
of parameters. Simple parameters are the total or the mean of a study variable,
or the ratio of the totals of two study variables. Different types of elements are
sometimes measured in the same survey, as when both individuals and households
are observed.

Many surveys are conducted periodically, for example monthly or yearly. As
a consequence, an important objective is to measure the change in the level of a
variable between two survey occasions. The objectives of estimation of change
and estimation of level often coexist in a survey, but they may require somewhat
different techniques.

A survey usually originates in an expressed need for information about a social
or economic issue, a need which existing data sources are incapable of filling. The
first step in the planning process is to determine the survey objectives as clearly
and unambiguously as possible. The next step, referred to as *survey design*, is to
develop the methodology for the survey.

Survey design involves making decisions on a number of future survey operations. The data collection method must be decided on, a questionnaire must be
designed and pretested, procedures must be set out for minimizing or controlling
response errors, the sampling method must be decided on, interviewers must be
selected and trained, questionnaires must be constructed and tested, techniques
for handling nonresponse must be agreed on, and procedures for tabulation and
analysis must be settled.

A survey will usually encounter various technical difficulties. No survey is
perfect in all regards. The statistics that result from the survey are not error-free.
The *frame* from which the sample is drawn is hardly ever perfect, so there will be
coverage errors. There will be *sampling error* whenever observation is limited

---

Estimation in Surveys with Nonresponse   C.-E. Särndal and S. Lundström
© 2005 John Wiley & Sons, Ltd

to a sample of elements, rather than to the entire population. No matter how carefully the survey is designed and conducted, some of the desired data will be missing, because of refusal to provide information or because contact cannot be established with a selected element. Since nonresponding elements may be systematically different (for example, have larger or smaller variable values, on average) from responding elements, there will be *nonresponse error*.

These three types of error – sampling error, nonresponse error and coverage error – are discussed at length in this book, especially the first two. It is true that a survey will usually also have other imperfections, such as measurement error and coding error. These errors are not discussed.

Subpopulations of interest are called *domains*. If the survey is required to give accurate information about many domains, a complete enumeration within these domains may become necessary, especially if they are small.

The survey planner is likely to first consider whether statistics derived from available *administrative registers* could satisfy the need for information. This avenue can be followed in countries well endowed with high-quality administrative registers. If not, a *census* (a complete enumeration of the population) may have to be conducted. If all domains of interest are at least moderately large, a *sample survey* may give statistics of sufficient accuracy.

These three different types of survey – based on administrative registers, census survey and sample survey – differ not only in the extent to which they can produce accurate information for domains, but also in other important respects. For example, sample surveys have the advantage of yielding diverse and timely data on specified variables, whereas statistics derived from administrative registers, although perhaps less expensive, may give information of limited relevance, because except in fortunate circumstances, available registers are not designed to meet specific information needs. On the other hand, a census might provide the desired information with great accuracy, but is very expensive to conduct. For a discussion of these issues, see Kish (1979).

Most of the issues raised in the following apply to all three types of survey. But most often, we will have in mind a sample survey. Therefore, the term 'survey' will usually refer to a 'sample survey'. We will now review some frequently used survey terminology.

A survey seeks to provide information about a *target population*. The delimitation of the target population must be clearly stated at the planning stage of the survey. The statistician's interest does not lie in publishing information about individual elements of the target population (such disclosure is often ruled out by law), but in measuring quantities (totals or functions of totals) for aggregates of elements, the whole population or domains. These targeted quantities are called *parameters* or *parameters of interest*. For example, three important objectives of a labour force survey (as conducted in most industrialized countries) are to obtain information about the number of unemployed, the number of employed and the unemployment rate. These are examples of parameters. The first two parameters are *population totals*. The third is a *ratio of population totals*, namely, the number of unemployed persons divided by the total number of persons in the labour force.

Examples of other population parameters are *population means* – for example, mean household income – and *regression coefficients*, say, the regression coefficient of income (dependent variable) regressed on number of years of formal education (independent variable), for a population of individuals. We can estimate any of these parameters with the aid of data on a sample of elements.

The sample is a selection from the *frame population*. The frame population is a list or other device that identifies and represents all elements that could possibly be drawn. Ideally, the frame population represents exactly the set of physically existing elements that make up the target population. In reality, the frame population and the target population differ more or less, as we discuss in more detail later.

Sampling design is used as a generic term for the (usually probabilistic) rule that governs the sample selection. Commonly used sampling designs are: simple random sampling (SI), stratified simple random sampling (STSI), cluster sampling, two-stage sampling, and probability-proportional-to-size (πps) sampling, of which Poisson sampling is a special case. With the possible exception of SI, these designs require planning before sampling can be carried out. STSI requires a set of well-defined strata. Cluster sampling requires a decision on what clusters to use. Sampling in two or more stages requires a clear definition of the first-stage sampling units, the second-stage units, and so on.

Every sampling design involves two other important general concepts: *inclusion probabilities* and *design weights*. The inclusion probability of an element is the known probability with which it is selected under the given sampling design. The design weight of an element is computed as the inverse of its inclusion probability, assumed to be greater than zero for all elements. Examples of designs where the inclusion probabilities are equal for all elements are SI and STSI with proportional allocation. Many sampling designs used in practice do not give the same inclusion probability to all population elements. In STSI, the inclusion probabilities are equal within strata, but they can differ widely between strata.

The inclusion probability can never exceed one. Consequently, a design weight is greater than or equal to one. The inclusion probability (and the design weight) is equal to one for an element that is selected with certainty. Many business surveys include a number of elements (usually very large elements) that are 'certainty elements'. These form a subgroup often called a *take-all stratum*.

A majority of the elements have inclusion probabilities strictly less than one. For example, in an STSI design, an element belonging to a stratum from which 200 elements are selected out of a total of 1600 has an inclusion probability equal to the sampling rate in the stratum, $200/1600 = 0.125$, and its design weight is $1/0.125 = 8$. One often heard interpretation is that 'an element with a design weight equal to 8 represents itself and seven other (non-sampled, non-observed) population elements as well'. When it comes to estimation, the observed value for this element is given the weight 8. Another stratum in the same survey may have 100 sampled elements out of a total of 200. Each element in this stratum has the inclusion probability $100/200 = 0.5$, and its design weight is then $1/0.5 = 2$.

STSI is a widely used design. It is very well suited for *surveys of individuals and households* in countries that can rely on a frame in the form of a total population register (see Example 2.1). Such a register lists the country's population and contains a number of variables suitable for forming strata, such as age, sex and geographical area. It is often of interest to measure households as well as individuals in the same survey. One way to obtain a sample of households from the sample of individuals is to identify the households to which the selected individuals belong. Household variables such as household expenditure can be observed, and variables on individuals, some or all of those residing in an identified household, can also be observed. We can obtain statistics on households as well as statistics on individuals.

The reverse order of selection is also possible. Practical considerations may necessitate drawing first a sample of households, with a specified sampling design, then selecting some or all of the individuals in the selected households. Again, both household variables and variables on individuals can be measured in the same survey. The selection of households can, for example, proceed by drawing a stratified sample of city blocks from a city map and then enumerating all the households in the selected city blocks.

In *business surveys*, the distribution of many variables of interest is highly skewed. The 'industry giants' account for a major share of the total for typical study variables related to production and output. The largest elements (enterprises) must be given a high inclusion probability (probability one or very near to one). Many business surveys use coordinated sampling for small enterprises to distribute the response burden. This entails some control over the frequency with which an enterprise is asked to provide information over a designated period of time, say a year. A number of countries have (to some extent different) systems for coordinated sampling. Statistics Sweden, for example, uses the system referred to as SAMU, described in Atmer *et al.* (1975). Another early reference for coordinated sampling is Brewer *et al.* (1972).

Coordinated sampling techniques are based on the concept of permanent random numbers: a uniformly distributed random number is attached at birth to a statistical element (an enterprise), and it remains with that element for the duration of its life; in that sense, it is permanent. The permanent random numbers play a crucial role in realizing both the desired inclusion probabilities and the desired degree of coordination of samples.

2.2. SOURCES OF ERROR IN A SURVEY

In this section we discuss frames, sampling and nonresponse. Figure 2.1 gives the background.

Coverage errors

We define the *target population* to be the set of elements that the survey aims to encompass at the point in time when the data are collected, by the completion of a

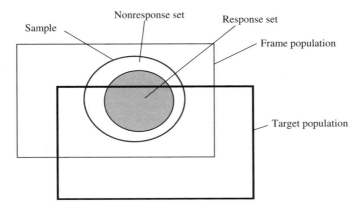

Figure 2.1   Representation of selected sample and response set, seen as subsets of the frame population, and their intersections with the target population.

questionnaire or in some other way. This point in time is called the *reference time point for the target population*. The sampling frame, on the other hand, is usually constructed at an earlier date, sometimes as much as 12 months earlier. This time point is referred to as the *reference time point for the frame population*. The lag between the two time points should be as short as possible, because the risk of coverage errors increases with the time lag. Three types of coverage error are commonly distinguished: *undercoverage, overcoverage* and *duplicate listings*. We comment on the first two of these in particular. As the name suggests, duplicate listings refer to errors occurring when a target population element is listed more than once in the frame.

Elements that are in the target population but not in the frame population constitute undercoverage. Especially in business surveys, a significant part of the undercoverage is made up of elements that are new to the target population and therefore absent from the frame population. These are commonly referred to as 'births'. Undercoverage may have other causes as well.

Elements that are in the frame population but not in the target population constitute overcoverage. Elements that have ceased to exist somewhere between the two reference time points can be a significant source of overcoverage. They are commonly referred to as 'deaths'.

It follows that undercoverage elements have zero probability of being selected for any sample drawn from the frame population. This is undesirable, because of the bias that can result if the study variable values for the undercoverage elements differ systematically from those of other population elements.

Bias from overcoverage can usually be avoided as long as it is possible to correctly identify the sample elements that belong to the overcoverage. But this is not always possible. For example, elements that are selected into the sample but become nonrespondents can often not be correctly classified as either 'in the target population' or 'in the overcoverage'. Biased estimates can be an undesirable consequence.

Although attempts may be made to minimize the lag between the frame population reference time point and the target population reference time point, the time lag is often considerable. It may be a practical necessity. One reason may be slow updating of the frame. As time goes by, events occur that motivate a change or update of the frame information. An example is a change in a variable value for an element existing in the frame, as when updated information is received about the number of employees or the gross business income of an enterprise. Such changes are sometimes recorded only with considerable delay.

It follows that the values recorded in the frame, at a given point in time and for a specific frame variable, may refer to different points in time for different elements. All elements are not necessarily updated at the same moment in time. This is not ideal, but it is a reality that has to be accepted.

Births and deaths are examples of events that need to be recorded. These events cause a change in the set of elements in the frame.

The frame population for a planned survey is sometimes created from a larger, more extensive collection or list of elements, each having recorded values for a number of variables. A frame population deemed appropriate for a particular survey may then be constructed from this larger collection, using some of the recorded variables to delimit the frame population. Imperfections in the recorded variable values, because of unequal reference times or other causes, may harm the effectiveness of the delimitation.

Imperfect frame variable values are undesirable for other reasons. For example, frame variables are often used before sampling to stratify the population and/or after sampling to poststratify the sample. Errors in the frame variables make these practices less efficient.

Example 2.1. The total population register

A total population register (TPR) exists in all Scandinavian countries (Denmark, Finland, Iceland, Norway and Sweden) and in the Netherlands. Such registers are being developed in other countries. A TPR aims at a complete listing of the country's population. Among the register variables recorded for every person are the person's unique personal identity number, name and address. This makes it possible to access, by mail or otherwise, every person for a wide range of surveys. The addresses are usually classified into administrative regions, such as counties and municipalities. The TPR may have every piece of real estate identified by coordinates, making it possible to construct regions other than counties and municipalities. Other important register variables are date of birth, sex, civil status (married, etc.), country of birth and taxable income. If we take Sweden as an example, information about births, deaths, immigration, emigration and changes of other register variables is received continuously by Statistics Sweden. The register can be kept almost perfectly up to date. Persons arriving from abroad and intending to stay at least 1 year are entered in the register, after the necessary permission has been granted. Since this is not instantaneous, the TPR at any given point in time have a minor degree of undercoverage, consisting of persons who properly belong to the Swedish population but who are not yet entered in the register. The information about births is almost error-free. There is also some overcoverage, that is, persons in the TPR but not (or no longer) in the Swedish population. This overcoverage, estimated to be around 0.4 % of the entire population, is made up essentially of persons who have emigrated without proper notification for removal from the TPR. Since the personal identity number is unique, duplicates do not occur in the TPR. For STSI, the sizes of strata based on variables such as sex, age group, civil status and many others are determined by a simple count in the TPR.   □

Example 2.2.   The business register

A business register (BR) that lists enterprises and work establishments (local units) is maintained by the statistical agencies in many countries. Although countries differ in the specifics, many principles are common or similar between countries. Let us again use Sweden as an illustration. Every week, Statistics Sweden's BR receives information from the National Tax Board about births and deaths of enterprises. The births fall into three categories: (i) pure births, that is, enterprises generated by new business activity; (ii) births occurring because of reorganization, as when an existing enterprise is split into several entities; (iii) births arising because of a registration of an enterprise under new legal form. A large percentage of the births may belong to (ii) and (iii). Among the BR variables, two important ones for an enterprise are the Standard Industrial Classification (SIC) code and the number of employees. The information for updating these variables comes from several different sources, specializing in different subsets of the BR. At any given point in time, the most recently recorded variable values do not necessarily refer to the same point in time.

Only at a limited number of times each year is the BR an ideal source for establishing a sampling frame. Those are the times just after the completion of an important updating activity. For updating, the BR unit at Statistics Sweden receives information partly from external sources, partly from its own specially designed surveys. At Statistics Sweden, new sampling frames for business surveys are produced in March, May, August and November, based on the BR as it exists at those different points in time. These frames are of good quality, without being perfect in every regard.   □

Sampling error and nonresponse error

When survey statisticians speak about *sampling error*, they mean the error caused by the fact that values of a study variable are solicited only for a sample of elements, not for the whole population. If the whole population were indeed observed, the sampling error would be zero. (There could be other errors, for example, measurement error and nonresponse error, but the sampling error would be zero.) This situation is exceptional. Statisticians often measure 'error' using the concept of variance. Hence, the sampling error is measured by the variance of the estimator in use, *assuming that there are no other errors*.

The usual technique for estimating a population total consists in summing appropriately weighted variable values for the responding elements in the sample. Different weighting systems can come into consideration.

We can use the design weights, given by inverting the inclusion probabilities. If all sampled elements respond (the case of complete response), this gives an unbiased estimator for the population total of the variable in question. This is known as the Horvitz–Thompson (HT) estimator and is discussed further in Section 4.2.

A more sophisticated (and usually more efficient) weighting is the one used in the generalized regression (GREG) estimator, discussed in Section 4.3. For complete response, a GREG point estimate of a population total is also a sum of weighted observed values, in such a way that the weight of an element is the product of the design weight and the *g-weight*. The latter, explained in more detail in Section 4.3, is computed with the aid of the available auxiliary information. *Poststratification* weights are a simple type of *g*-weight familiar to many statisticians.

We have already stated that many parameters of interest in a survey are more complex than a population total or a population mean of a variable. A number of parameters of interest are expressible as a function of two or more population totals. A simple principle is used for estimating a function of totals: replace each unknown population total by its HT estimator or by a GREG estimator. For example, to estimate wages per hour worked (in an industry, for example), we compute the (HT or GREG) estimate of total wages and divide it by the (HT or GREG) estimate of total hours worked.

The *variance* of an estimator is the average of the square of the deviation of the estimator from its central value (mean). This average is with respect to all possible samples that can be drawn using the given sampling design. Since each of these samples has a known probability, determined by the sampling design, we can derive a formula for the variance. It is important to note that variance is measured as 'variability over all possible samples'. But in practice we never draw all possible samples; we draw just one single sample. Nevertheless, we seek to quantify the variance by a computation based on the data we do have. This is what *variance estimation* sets out to do.

The variance of an estimator is an unknown quantity. It depends on data for the whole population. We *estimate* the variance with the aid of the values observed for the sampled elements. When this is possible the sampling design is said to be *measurable*. The objective is to do this so that the variance estimator is (almost) unbiased.

The estimated standard deviation is defined as the square root of the estimated variance of the estimator. It is used in confidence interval calculation. The familiar procedure for obtaining a confidence interval at (roughly) the 95 % level is to compute the endpoints of the interval as: point estimate plus or minus 1.96 times the estimated standard deviation. It is important that an unbiased point estimate is used. If not, the interval will be off-centre, and it will not carry the desired 95 % confidence.

In the last two decades, computer software has been designed in order to handle estimation in surveys. Such software will typically handle weight computation, point estimation and the corresponding variance estimation. Some software handles the computation of weights for survey data affected by nonresponse. Software of this general kind, with more or less extensive features, has been produced in a number of countries. It includes BASCULA from the Netherlands (see Nieuwenbrook and Boonstra, 2002), CALMAR and POULPE from France (see Deville *et al.*, 1993; Caron *et al.*, 1998; Caron, 1998), CLAN97 from Sweden (see Andersson and Nordberg, 1998) and GES from Canada (see Estevao *et al.*, 1995). A number of other countries could be mentioned, including Belgium, Italy and the United States. Software development is ongoing. Existing software is updated. New products appear. For particulars about a given software, the reader is referred to the references mentioned and to related material.

Considerable resources are spent in many countries on improving data collection procedures, so as to keep the *nonresponse* as low as possible. Nevertheless, once data collection is concluded, one has to accept some, perhaps even considerable, nonresponse. It causes *nonresponse error* in the estimates. A 20 % nonresponse rate is common, and in many surveys it is much higher. Nonresponse rates are on the increase in many surveys and many countries.

For example, for the year 2000, Statistics Sweden reports nonresponse rates between 25 % and well over 30 % for several surveys, including those relating to consumer buying expectations, income distribution, national and international road goods transport, and transition from upper secondary school to higher education. The last mentioned survey is analysed in detail in Section 14.4.

Internationally, one finds that response rates vary greatly between countries as well as between types of survey. An analysis of patterns in household survey nonresponse is reported in de Leeuw and de Heer (2002). They collected time series data for 10 different surveys from producers of official statistics in 16 developed countries. Not all countries had provided data for each survey. All 16 countries had supplied data for the labour force survey, a majority of them for the family expenditures survey, and some of them for surveys related to health, to travel and to income. The length of the time series varied between countries and between surveys.

The data were carefully analysed by multilevel logistic models. Three dependent variables were response rate, noncontact rate and refusal rate. The latter two are the traditionally recognized major components of nonresponse. The explanatory variables were year (as a fixed effect) and country, survey and year (as random effects). The main results are summarized in the following points. The effects mentioned were found statistically significant.

- Response rates have declined over the years. Countries as well as surveys differ with regard to response rates.
- Noncontact rates have increased over the years, and they differ between countries.
- Refusal rates have increased over the years, and they vary between countries and between surveys.

The noncontact rates were not found to differ significantly between surveys, something which may be explained by a similarity of field work between surveys in one and the same country. The overall noncontact rate was estimated at 6 %, the overall refusal rate at 9 %. The overall increase per year in the noncontact rate was estimated at 0.2 %, that of the overall noncontact rate at 0.3 %.

It is worth noting that the surveys analysed are still, at this point in time, well respected, and generally viewed as being in the public interest. Comparatively speaking, the response rates in those surveys are good. Considerably lower response rates are encountered in many other surveys, in national statistical agencies as in private survey institutes. The theory and the methods in this book, and in important references that we have relied on, may be applied to surveys with 'typical nonresponse rates', considering what the conditions are at this point in time. Nevertheless, the methods in this book are not a panacea; the objective of a survey should always be to achieve as favourable a response rate as possible, under the given constraints of cost and other matters.

For surveys on individuals, the existing vast literature contains much information about nonresponse distributed with respect to basic variables such as age, sex, and region. Experience gathered from these *nonresponse analyses* shows that, for surveys on individuals, lower response rates are usually expected for metropolitan residents, single persons, members of childless households, older persons, divorced or widowed persons, persons with lower educational attainment, and self-employed persons; see Groves and Couper (1998), Holt and Elliot (1991), Lindström (1983). A review of issues in unit nonresponse and item nonresponse for business surveys is given in Willimack *et al.* (2002).

Since variables such as age, sex and region often covary with many social survey study variables, the nonresponding elements are likely to be atypical with respect to these variables. This causes *nonresponse bias* in the estimates, unless proper action can be taken at the estimation stage. A distinct possibility is that some, or even considerable, bias remains in the estimates even after a carefully executed estimation.

Another effect of nonresponse is an *increase in the variance* of estimates, because the effective sample size is reduced. If an increased variance were the only problem, it could be fairly easily counteracted by some degree of 'oversampling', so that the sample size is fixed at the design stage at an appropriately 'higher than normal' rate. The only negative effect of nonresponse might then be some increase in administrative burden and in data collection cost, through higher postage fees, for example.

But it is the nonresponse bias that constitutes by far the most important obstacle to correct statistical conclusions in a survey. Compared to the bias, the increased variance must be considered only a minor disturbance. In the presence of a significant bias, a computed confidence interval will be centred on the wrong value and thus misleading. It does not carry the level of confidence required.

A related disturbance, in the case of a stratified design, is that if the sample is allocated to strata in an optimal fashion, the nonresponse that occurs may cause the responding elements to distribute themselves in a far from optimal manner over the strata. The normally important gains of stratification can be severely compromised by the nonresponse.

Table 2.1   Illustration of data present (marked ×) and data missing (marked nr) in a hypothetical data collection.

Identity	Register variables		Questionnaire variables		
	1	2	1	2	3
1	×	×	×	×	×
2	×	×	×	×	nr
3	×	×	×	nr	×
4	×	×	×	×	×
5	×	×	×	×	×
6	×	×	nr	×	nr
7	×	×	nr	nr	nr
8	×	×	nr	nr	nr

A survey may contain many study variables. In some surveys, it is possible to obtain data on some of these variables from available registers. We call them *register variables*. These sources of information will typically show data on all variables and all elements; no data are missing. However, for the other study variables, data are solicited from a sample selection of elements by telephone or mail or electronically, using a questionnaire or similar data transfer device. The data for these *questionnaire variables* are invariably affected, more or less, by nonresponse. In the following, 'study variable' will ordinarily mean 'questionnaire variable.' It is customary to distinguish two types of nonresponse: *unit nonresponse* and *item nonresponse*. We shall use the following definitions. A unit nonresponse element is one for which information is missing on all the questionnaire variables. An item nonresponse element is one for which information is missing on at least one, but not all, of the questionnaire variables. The set of elements with a recorded response on at least one questionnaire item will be called the *response set*. These concepts are illustrated by the following example.

Example 2.3.   Unit nonresponse, item nonresponse and response set

Table 2.1 illustrates the result of a (hypothetical) data collection in a survey with eight sampled elements. The symbol × indicates a presence of data, and nr indicates that data are missing. Elements 1, 4 and 5 have complete response. Elements 2, 3 and 6 have some item nonresponse, that is, one or more values are missing among the three questionnaire variables. Although all eight sample elements have data for the two register variables, we shall say that elements 7 and 8 constitute the unit nonresponse, because both of these have no response at all to the questionnaire part of the survey. Elements 1–6, all of which have values recorded for at least one questionnaire item, form the response set in this example.   □

CHAPTER 3

General Principles to Assist Estimation

3.1. INTRODUCTION

Estimation in the presence of nonresponse is a collective term for the methods used to produce statistics (that is, to make estimates) when the survey is affected by nonresponse. We are at a point where the estimation must go ahead without some of the data that one had hoped for; one must do the best possible with the data at hand. The data file that we have to work with specifies observed values on a given study variable not for all those in the selected sample, but 'only' for those who responded to that variable. There are missing values, more so for some study variables than for others. Two types of methods are at our disposal, *weighting* and *imputation*.

Weighting entails computing appropriate weights to be applied to the observed values for respondents. Since observations were lost by nonresponse, weighting will have to compensate for this. The weights will be increased, on average, compared to a complete response. In this book, weighting is systematically developed by a general approach, the *calibration approach*, in Chapters 6–11. In particular, the question of nonresponse bias and how to reduce it is an important topic discussed in Chapters 9 and 10. An appealing property of the calibration approach is that it includes as special cases many of the 'standard methods' seen in the literature. Chapter 7 brings out this point in more detail.

Imputation entails replacing missing values by proxy values. A common approach is to use imputation for item nonresponse only and then treat unit nonresponse by weighting. We call this the *combined approach*. It is explored in Chapters 12 and 13. There exist many methods to construct imputed values. A number of those in common use are discussed in Chapter 12.

Example 3.1. Illustration of sampling error and nonresponse error, assuming a perfect frame

In order to illustrate certain nonresponse adjustment procedures we constructed an artificial population of size $N = 34\,478$ consisting of $17\,062$ men and $17\,416$ women. The data came from Statistics Sweden's Income Distribution Survey

1999 and the constructed study variable value y_k represents the sum of total yearly earned and capital income of person k. The mean value of y_k was SEK 196 592 for men and SEK 135 689 for women.

Suppose we want to estimate total population income, that is, the total

$$Y = y_1 + y_2 + \cdots + y_N = \sum_{k=1}^{N} y_k = 571 \times 10^7$$

of the $N = 34\,478$ values y_k, under the following survey conditions. We draw an SI sample of size 400, then let a response mechanism operate in such a way that all males respond with probability 0.5 and all females with probability 0.9. (Since this is an experiment, we can choose whatever response distribution we like. In a real survey, we are less fortunate: the response distribution is unknown.) The response set has a random size varying around an expected value of $400 \times (17\,062 \times 0.5 + 17\,416 \times 0.9)/34\,478 = 281$ and will tend to strongly overrepresent women, compared to the more natural male-to-female distribution found in the desired SI sample. Assume that we estimate the total income by the expansion estimator $\hat{Y}_{\text{EXP}} = N\bar{y}_r$, so called because it expands the mean $\bar{y}_r$ of the respondent y_k-values to the population level through a multiplication by N. This estimator, given by (7.2), implies that we treat the response set as a simple random selection from the population. This is incorrect, under the response distribution that we have specified. Consequently, the estimates derived with the formula $N\bar{y}_r$ will be biased. To show this, we drew a total of 100 SI samples, and for each of these, a response set was obtained by the response mechanism mentioned. The results are given in Figure 3.1. The horizontal axis represents the numbering, from 1 to 100, of the response sets, and the vertical axis represents the estimate of total population income multiplied by 10^{-7}, that is, $\hat{Y}_{\text{EXP}} \times 10^{-7}$. The horizontal straight line indicates the target value 571. The figure shows the variability of the estimates and, more specifically, it shows the bias caused by nonresponse.

The majority (although not all) of the 100 estimates, as well as their mean, fall considerably below the target value. The distinctly negative bias is caused by a nonresponse pattern in which men respond less than women. The bias cannot be estimated from a single sample, which is all we have in a real survey. The variance, illustrated by the fluctuation of the 100 estimates around their average,

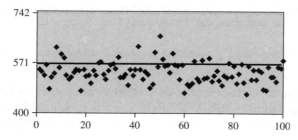

Figure 3.1 One hundred expansion estimates of the population total. Each estimate is multiplied by 10^{-7}.

is the sum of a sampling variance and a nonresponse variance, as discussed in Section 5.2. The nonresponse causes not only a bias but also an increase in variance. The sampling variance alone is smaller than the variance indicated by the figure.

We can trace the failure of the estimator $N\bar{y}_r$ to inefficient weighting. The values y_k are equally weighted. Every element receives the weight N/m, where m is the number of respondents. But men are underrepresented among the respondents. For an improved estimation, men should be given higher weights than women. We can specify such improved weights with the use of auxiliary information. How this can be done is shown in Example 3.2.   □

3.2. THE IMPORTANCE OF AUXILIARY INFORMATION

The key to successful weighting for nonresponse lies in the use of powerful auxiliary information. This will reduce both the nonresponse bias and the variance.

Register variables play an important role in many surveys. They are used in creating an appropriate sampling design and/or in the computation of the survey estimates. In both uses, the register variables can be called *auxiliary variables*, because they assist and improve the procedures. Most often, as usually in this book, the term 'auxiliary variable' refers to a variable used at the estimation stage to improve on less efficient estimators.

Register variables are frequently used to construct the strata for stratified sampling designs. Such designs allow us to achieve a desired precision in estimates for the whole population and for particularly important domains (subpopulations). Each important domain should be designated as a separate stratum. In other surveys, particularly in business surveys, a register variable may be used as the 'size variable' necessary for constructing a πps design, in the manner discussed in many papers, textbooks and in Section 4.2 of this book.

Terms frequently used in the following are *auxiliary variable*, *auxiliary vector*, *auxiliary information* and *auxiliary population total(s)*. We use these terms in the following manner. The minimum requirement to qualify as an auxiliary variable is that the values of the variable are available for every *sampled* element (that is, for both responding and nonresponding elements). Better still, the auxiliary variable values may be available for every element in the whole population. An example of this is when statistical agencies use auxiliary variables specified in different existing registers, for all elements in the population. By simply summing the variable values given in the register, we obtain the population total of that auxiliary variable. This information is needed in some of the estimators to be considered.

An auxiliary vector is made up of one or more auxiliary variables. There are two important steps in building the auxiliary vector that will be ultimately used in the estimation: making an inventory of potential auxiliary variables; and selecting the most suitable of these variables and preparing them for entry into the auxiliary vector. The auxiliary variables deemed potentially useful for the estimation may come from several registers such that a linking of elements is possible with the aid of identifier keys. A rather long list of potential variables may result from this scrutiny. Then comes the procedure by which we arrive at the final form of the auxiliary vector to be used in the estimation. This process may require considerable reflection and analysis. Decisions to be taken include the selection of variables from the available larger set, the setting of appropriate group boundaries for converting a quantitative variable into a categorical variable, and the fixing of rules for collapsing very small groups into larger groups.

Note that when register variables are used in the construction of the sampling design, their values must be known for every element in the population. A case in point is when the strata are constructed for use in stratified sampling. By contrast, when auxiliary variables are used at the estimation stage, such detailed information may not be necessary. It may suffice to know, from some exterior but accurate source, the population total for each auxiliary variable, while the knowledge of individual variable values is limited to the responding elements.

The following simple example illustrates how nonresponse bias can be reduced by incorporating relevant auxiliary information in the estimation procedure.

Example 3.2.   Reducing the nonresponse bias by the use of auxiliary information

In Example 3.1, the expansion estimator $N\bar{y}_r$ was found to have an unacceptably large bias under SI. The bias is explained in major part by an absence of auxiliary information. To continue the example, suppose that we can use sex as an auxiliary variable. That is, we know the population frequency of men, N_1, and of women, N_2. This information allows us to use the population weighting adjustment estimator $\hat{Y}_{\mathrm{PWA}}$, given by (7.5). Here there are $P = 2$ classes, men and women. (Some call it the poststratified estimator; the terminology is discussed in Section 7.3.) It calls for the weight N_1/m_1 to be applied to every one of the m_1 responding men and the weight N_2/m_2 to every one of the m_2 responding women. We computed this alternative estimator for each of the 100 response sets in Example 3.1. The results are shown in Figure 3.2. As in Example 3.1, the horizontal axis represents the numbering, from 1 to 100, of the response sets. The vertical axis represents estimated total population income multiplied by 10^{-7}, that is, $\hat{Y}_{\mathrm{PWA}} \times 10^{-7}$. The horizontal straight line indicates the target value $Y \times 10^{-7} = 571$.

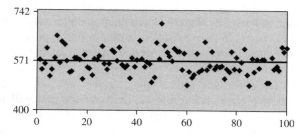

Figure 3.2   One hundred estimates, by population weighting adjustment, of the population total. Each estimate is multiplied by 10^{-7}.

There is a striking improvement compared with Figure 3.1. By visual inspection alone, the mean of the 100 poststratified estimates is now seen to be very close to the target value. (It can be shown theoretically that the bias is very nearly zero.) □

We concentrate in this book on efficient estimation, rather than on sampling design or other survey operations. However, let us consider one example of how nonresponse bias can be controlled by building auxiliary information into the sampling design.

Example 3.3. Reduction of nonresponse bias by the use of auxiliary information in the sampling design

A commonly used procedure has the following two features: (i) stratified simple random sampling, followed by (ii) straight expansion within each stratum, using the inverse of the stratum response fraction. The term 'straight expansion within strata' is appropriate for this method. Because of its simplicity it is convenient in routine statistics production. But in practice, there is often little or no effort made to verify whether the assumption behind the method is satisfied, namely, that every element within a given stratum responds with the same probability.

Suppose we follow this procedure for the population discussed in Examples 3.1 and 3.2. Let there be two strata, men and women, and nonresponse adjustment by straight expansion in each stratum. Although we do not present a new diagram, it can be shown (as Chapter 9 does) that the nonresponse bias is nearly the same as in the weighting class procedure in Example 3.2, and thus considerably reduced compared to Example 3.1. This illustrates that entering the auxiliary information into the sampling design can also help in reducing the nonresponse bias.

As mentioned previously, important domains of interest are often designated as strata for stratified sampling. This permits the total available survey resources to be allocated in such a way that every important domain is sufficiently represented to achieve a desired level of precision. One may decide to overrepresent smallish domains, compared to larger ones. For instance, if geographic areas are important domains of interest, then the strata should correspond to these areas.

It should be noted that although the procedure defined by (i) and (ii) is very common in surveys, it amounts to a restricted, or unimaginative, use of auxiliary information. If applied in a mechanical fashion, the procedure is oblivious to more effective options that could perhaps be chosen after making inventory and constructive use of other auxiliary information. □

Virtually all techniques described in this book use auxiliary information in one form or another. Powerful auxiliary information is particularly important in treating nonresponse, because such information lends strength to reweighting and imputation procedures, thereby reducing several errors: sampling error, nonresponse error and coverage error.

Example 3.4.   Examples of reduced nonresponse bias as a result of identifying new powerful auxiliary variables

That a search for more powerful auxiliary information can significantly improve the estimates is illustrated by recent developments in the Labour Force Survey (LFS) in Finland and in Sweden. The old method, in both countries, involved expanding the weights by the inverse of the response rate within groups based on age, sex and region. In both countries it was found that the inclusion of a new register variable into the auxiliary vector improved the estimation considerably. In its simplest form this variable is dichotomous, with the value 1 for a person registered in the country's job seekers' register, and 0 otherwise. It is quite understandably highly correlated with 'being unemployed'. For the estimation, the register is matched with the LFS sample with the aid of the unique personal identification number. The estimated number of unemployed, most likely too low under the old design, became significantly higher after inclusion of the job seeker variable into the auxiliary vector. This was revealed by a comparison of the old estimation method with the new estimator. This important change in the level of the estimates is most likely attributable to a considerable reduction of the nonresponse bias. These developments are reported in Djerf (1997, 2000) and Hörngren (1992).   □

3.3. DESIRABLE FEATURES OF AN AUXILIARY VECTOR

The examples in Section 3.2 involved simple special cases of the calibration estimator. We turn now to the principles that should guide the selection of auxiliary variables in more general uses of the calibration approach to weighting. The approach is described starting in Chapter 6. Once the auxiliary vector has been finally fixed, the calibrated weights can be computed in a straightforward manner, using existing software such as CLAN97. To compose the best possible auxiliary vector is an important part of the estimation process.

The auxiliary vector should be one that:

  (i)   shows clear evidence of explaining the response propensity;
 (ii)   shows clear evidence of explaining the main study variables;
(iii)   identifies, or comes close to identifying, the most important domains.

These desirable properties require some comment. They are stated here in a rather informal manner. In Chapter 10, they are given more precise formulations, with a background in the theory developed from Chapter 6 onwards. Fulfilling (i) contributes to reducing the nonresponse bias in the estimates for all study variables. Fulfilling (ii) contributes to reducing the nonresponse bias in the estimates for the main study variables. It will also work in the direction of reducing the variance in the estimates for these variables. The principal effect of satisfying (iii) is a reduction of both bias and variance in the estimates for the most important domains. That the auxiliary vector should identify important domains means that the auxiliary totals produced from the vector should be totals pertaining to the domain itself.

The following example illustrates how one may reason in practice to fulfil the principles. For this purpose we use Statistics Sweden's Survey on Life and Health. Several registers on individuals are available; together they provide a rich source of potential auxiliary variables. The example illustrates how one may reason in making choices from this wide base of information and end up with a suitable auxiliary vector for the calibration.

Example 3.5. The Survey on Life and Health (Liv och Hälsa)

The population consists of persons aged 18–79 in the county of Södermanland, Sweden. As the name suggests, the study variables measure different aspects of life and health. The survey managers view some of these variables as more important than others; they can be designated as the main study variables. The frame population, as determined by the TPR (see Example 2.1), was stratified by municipality. (There are nine municipalities.) The total sample was allocated to the strata so as to meet specified precision requirements for each municipality. Within each stratum an SI was drawn. At the end of the data collection stage, the nonresponse rate was found to be 34.4 %. This very high rate could have caused a substantial nonresponse bias if no effort had been made to reduce the bias. Fortunately, strong auxiliary information was available. Through the use of the calibration approach, it became possible to construct an estimator believed to protect against nonresponse bias.

Two sources of auxiliary information were used: the TPR and the Register of Education. As the result of an inventory, six prospective auxiliary variables, all of them categorical, were retained: sex (male; female), age group (four classes), country of birth (the Scandinavian countries; other), income group (three classes), civil status (married; other) and education level (three classes).

Two different analyses were carried out to identify the most suitable auxiliary variables. More specifically, the objective was to see which, if any, of the six variables were particularly relevant for explaining the response pattern and for explaining the main study variables.

A *nonresponse analysis* was carried out, consisting of a computation of the response rates, weighted by strata, in the different classes defined for each of the six variables. The results are given in Tables 3.1–3.6. For every one of the

Table 3.1   Response rate (%) by sex.

Sex	Male	Female
Response rate (%)	60.1	71.2

Table 3.2   Response rate (%) by age group.

Age group	18–34	35–49	50–64	65–79
Response rate (%)	54.9	61.0	72.5	78.2

Table 3.3 Response rate (%) by country of birth.

Country of birth	Scandinavian countries	Other
Response rate (%)	66.7	50.8

Table 3.4 Response rate (%) by income group.

Income class (in thousands of SEK)	0–149	150–299	300+
Response rate (%)	60.8	70.0	70.2

Table 3.5 Response rate (%) by civil status.

Civil status	Married	Other
Response rate (%)	72.7	58.7

Table 3.6 Response rate (%) by education level.

Education level	Level 1	Level 2	Level 3
Response rate (%)	63.7	65.4	75.6

six prospective categorical variables, the response rates differ considerably for the different classes of a variable. Thus, we expect that all six may contribute to explaining the response propensity. The response rates are close in the last two income groups and in the first two educational groups, suggesting that they should perhaps be collapsed. However, maintaining all the groups may contribute to satisfying principle (ii), so no collapsing is undertaken at this stage.

Those in charge of the survey were asked by the methodologists to identify the most important study variables. Several variables were mentioned, among them several dichotomous variables. A dichotomous variable has the value 1 for a person who has the attribute in question and 0 otherwise. Four important dichotomous study variables were defined by the following attributes: (a) poor health; (b) avoiding staying outdoors after dark (for fear of attack); (c) difficulties with regard to housing; (d) poor personal finances. An analysis was carried out for each of these four attributes to see how well it could be explained by the six prospective auxiliary variables.

For each attribute, the estimated proportion of persons possessing the attribute was computed, and this was done for every class defined by a categorical auxiliary variable. These estimates, given in Tables 3.7–3.12, were computed by straight

Table 3.7 Estimated proportion (%) of individuals having attributes (a)–(d), by sex.

Attribute	Male	Female
(a)	7.5	8.9
(b)	7.8	21.1
(c)	2.6	2.4
(d)	19.6	19.8

Table 3.8 Estimated proportion (%) of individuals having attributes (a)–(d), by age group.

Attribute	18–34	35–49	50–64	65–79
(a)	4.3	6.6	10.6	10.9
(b)	11.8	11.4	14.3	23.4
(c)	5.9	2.8	1.0	0.8
(d)	31.0	26.6	12.5	9.6

Table 3.9 Estimated proportion (%) of individuals having attributes (a)–(d), by country of birth.

Attribute	Scandinavian countries	Other
(a)	8.0	11.7
(b)	14.7	18.3
(c)	2.4	4.2
(d)	19.2	28.5

Table 3.10 Estimated proportion (%) of individuals having attributes (a)–(d), by income group (in thousands of SEK).

Attribute	0–149	150–299	300+
(a)	10.0	7.2	4.0
(b)	18.6	12.6	8.1
(c)	3.8	1.5	1.0
(d)	25.3	16.5	6.9

expansion within strata, as explained in Example 3.3. The tables hint that the six prospective auxiliary variables explain the study variables to somewhat different degrees. Most of the six appear to be reasonably good explanatory variables. Sex and civil status seem weaker than the other four, at least for some of the four study variables.

The prospective auxiliary variables are to some extent intercorrelated, so if all of them were retained as input into the calibration, some of the information would

Table 3.11   Estimated proportion (%) of individuals having attributes (a)–(d), by civil status.

Attribute	Married	Other
(a)	8.2	8.2
(b)	13.8	16.3
(c)	1.1	4.3
(d)	14.1	26.5

Table 3.12   Estimated proportion (%) of individuals having attributes (a)–(d), by education level.

Attribute	Level 1	Level 2	Level 3
(a)	10.5	7.3	4.6
(b)	19.1	12.6	12.9
(c)	1.7	3.2	1.8
(d)	17.5	21.6	16.8

be redundant. For a few responding persons, this could lead to abnormally high or low, even negative, weights. An increase in variance may be an undesirable consequence. Groups that are too small may produce the same effect. Therefore, a prudent approach is always to analyse the distribution of the calibrated weights, and in particular to watch out for extreme weights. If a few occur, trimming should be used.

In this particular survey the decision was finally reached to use an auxiliary vector composed of five categorical variables: municipality, sex, age group, country of birth and education level. Out of these, municipality is also used as a stratification variable. This does not preclude its inclusion into the auxiliary vector; in fact, in order to ensure that municipality estimates add up correctly, it must be included.

The domains considered to be of principal interest in this survey are those determined by the cross-classification sex by age group by municipality, which we represent by the expression *Sex × Age group × Municipality*. In order to satisfy principle (iii), the auxiliary vector should thus include *Sex × Age group × Municipality*. If the vector is taken as *Sex × Age group × Municipality* and nothing else, the calibration estimator becomes a population weighting adjustment estimator with $2 \times 4 \times 9$ classes. However, the decision was made to also include the auxiliary variables country of birth and education level. If this were to be done in the most detailed manner, we would have a complete cross-classification of all five variables. But then some of the five-dimensional cells may contain very few observations, or none at all. An increase in variance might be an undesirable consequence. Therefore, country of birth and education level were included as separate variables. The final auxiliary vector can then be represented as *Sex × Age group × Municipality + Country of birth + Education level*. Its dimension is $(2 \times 4 \times 9) + 2 + 3$. The calibrated weights computed by means of this auxiliary vector have the following properties. When applied to the auxiliary vector, they will exactly reproduce: (a) the known population counts for the cells determined by *Sex* × *Age group* × *Municipality*; (b) the known marginal counts in the population for country of birth; (c) the known marginal counts in the population for education level. In this survey, the calibration estimator and the corresponding variance estimator were calculated, for all domains of interest, by CLAN97.

□

Essentially all the estimation methods reviewed in this book revolve around different uses of auxiliary information. The calibration approach to weighting is introduced in Chapter 6. The technique is very general. As Chapter 7 points out, many 'conventional methods' are obtained as special cases, when the auxiliary vector has certain simple forms. Chapter 8 deals with extensions of the technique to cases where auxiliary information exists both at the level of the population and at the level of the sample. An issue of major importance is the bias that may remain in a calibration estimator even after a carefully executed weighting. The bias is analysed in Chapter 9, and, in consequence, Chapter 10 deals with methods to identify the most relevant auxiliary variables with a view to controlling the bias. Chapter 11 examines variance estimation for the calibration estimator, and Chapters 12 and 13 are devoted to imputation techniques. Finally, Chapter 14 presents extensions of the calibration approach to situations with imperfect frames.

Before commencing the systematic development of the calibration approach in Chapter 6, we need to recall how auxiliary information is taken into account in the case of complete response. Even though complete response practically never occurs in practice, it is a limiting case covered by the calibration approach when the nonresponse is reduced to nil. Chapter 4 therefore presents some necessary background.

CHAPTER 4

The Use of Auxiliary Information under Ideal Conditions

4.1. INTRODUCTION

Both nonresponse and frame imperfections are normal although undesirable features of any survey. Without them the quality of the statistics (the accuracy of the estimates) would generally be better. Neither of the two nuisance factors can be eliminated, at least not fully, at the *design stage* of the survey. We need techniques for dealing with them at the *estimation stage*.

This chapter contains a brief summary of the bases for estimation with auxiliary information under the 'ideal conditions' where the survey has complete response. The concepts of auxiliary variables and auxiliary information are central. They are used in this chapter, and then extended in Chapters 5–13 to handle estimation in the presence of nonresponse. In Chapter 14 we further extend the discussion to also deal with frame imperfections.

To fix ideas we introduce some notation. Consider the finite population of N elements $U = \{1, \ldots, k, \ldots, N\}$, called the *target population*. The letter k represents the kth element of the population, a physically existing element on which we can make measurements or observations. We wish to estimate the total

$$Y = \sum_{U} y_k \tag{4.1}$$

where y_k is the value of the study variable, y, for the kth element. (If A is any set of elements, $A \subseteq U$, we write the sum $\sum_{k \in A} y_k$ for simplicity as $\sum_A y_k$.)

The study variable can be continuous, or essentially so, as in situations where y_k stands for 'income' or 'energy production' of enterprise k. But it needs to be emphasized that in many important applications, y is a categorical study value, for example a dichotomous variable such that $y_k = 1$ if k has the attribute 'unemployed' and $y_k = 0$ otherwise, in which case $Y = \sum_{U} y_k$ is the number of unemployed in the population.

The values $y_1, y_2, \ldots, y_N$ are fixed, nonrandom real numbers. The random character comes, in this chapter, from the randomized sample selection and nothing else.

Estimation in Surveys with Nonresponse   C.-E. Särndal and S. Lundström
© 2005 John Wiley & Sons, Ltd

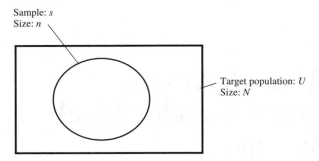

Figure 4.1 Representation of sample selection from the target population. No nonresponse.

Let s be a probability sample of size n, drawn from the target population U (see Figure 4.1) with the known probability $p(s)$. (Here, the frame population agrees with the target population.) The inclusion probabilities, known for all $k \in U$, are $\pi_k = \sum_{s \ni k} p(s) = \Pr(k \in s)$. Here $s \ni k$ indicates a summation over all the samples s which include element k. We assume that the design is such that $\pi_k > 0$ for all elements k. Let $d_k = 1/\pi_k$ denote the *design weight* of element k. The design weights are important for computing point estimators.

For computation of variance estimates we need, in addition to the π_k, to consider second-order inclusion probabilities. We denote by $\pi_{k\ell}$ the known probability that both k and ℓ are included in the sample, that is, $\pi_{k\ell} = \sum_{s \supset \{k,\ell\}} p(s) = \Pr(k \& \ell \in s)$. Here, $s \supset \{k, \ell\}$ indicates a summation over all the samples s which include both k and ℓ. The corresponding weight is denoted $d_{k\ell} = 1/\pi_{k\ell}$. The $d_{k\ell}$ are assumed to be defined for all $k \in U$ and $\ell \in U$. Note that if $k = \ell$, then $\pi_{k\ell} = \pi_{kk} = \pi_k$ and $d_{k\ell} = d_{kk} = d_k = 1/\pi_k$. The weights d_k and $d_{k\ell}$ are thus fully determined by the sampling design $p(s)$.

The designs commonly used are discussed in detail in basic sampling texts. Among these are SI and STSI. Another important category of designs are those that use πps sampling. The following examples illustrate the inclusion probabilities and the weights for STSI and πps sampling.

Example 4.1.   Design weights under the STSI design

Consider a STSI design with H strata indexed $h = 1, \ldots, H$ and such that n_h elements are selected by SI from N_h in stratum h. Then the design weights needed for the point estimation are $d_k = N_h/n_h$ for all k in stratum h. The weights $d_{k\ell}$ needed for the variance estimation are of three types:

$$d_{k\ell} = \begin{cases} d_{kk} = d_k = \dfrac{N_h}{n_h} & \text{if } k = \ell \text{ is in stratum } h, \\[2ex] \dfrac{N_h}{n_h}\dfrac{(N_h - 1)}{(n_h - 1)} & \text{if both } k \text{ and } \ell \text{ are in stratum } h, \\[2ex] \dfrac{N_h}{n_h}\dfrac{N_{h'}}{n_{h'}} & \text{if } k \text{ and } \ell \text{ are in different strata, } h \text{ and } h'. \quad \square \end{cases}$$

Example 4.2.   Design weights under πps sampling

Let z_k be a measure of the size of element k, known and positive for every $k \in U$. The number of employed is a typical size measure if the elements are business firms. A πps design is one whose first-order inclusion probabilities are given by

$$\pi_k = \frac{n z_k}{\sum_U z_k} \tag{4.2}$$

where n is the expected sample size, fixed in advance. For given values $z_1, \ldots, z_N$ there exist many such designs. One is *Poisson sampling*: a Bernoulli experiment is carried out for every $k \in U$, such that k is included in the sample if $u_k \leq n z_k / \sum_U z_k$, where u_k is a realization of $Unif(0,1)$, the random variable with a uniform distribution on the unit interval. The realizations $u_1, \ldots, u_N$ are assumed independent. The sample size is random with expected value equal to n.

The random sample size of Poisson sampling is viewed by some as undesirable. An interesting way to avoid this perceived drawback and still come very close to achieving the desired inclusion probability (4.2) is *order sampling*; see Rosén (1997a, 1997b). A particular case of order sampling is *Pareto sampling*, carried out as follows:

1   For every $k \in U$, generate u_k as a realization of $Unif(0,1)$, and compute

$$q_k = \frac{u_k(1 - \pi_k)}{(1 - u_k)\pi_k}$$

with π_k computed by (4.2).
2   Define the sample s to consist of those n elements that correspond to the n smallest values q_k.

The sample size is thus exactly equal to n. That the inclusion probabilities generated by the procedure composed of steps 1 and 2 come very close to satisfying (4.2) is not a trivial conclusion; the reader is referred to Rosén (1997a, 1997b) and references given there.   □

Both the sampling design and the point estimators are usually constructed with the aid of auxiliary information existing about the elements $k \in U$. Different information may be used for the sampling design and for the point estimation, but nothing prevents the use of some of the information at both stages.

The computational load in a survey is increased when estimates are needed not only for the whole population but also for a perhaps considerable number of subpopulations as well. They are called *domains of study* or *domains of interest* or simply *domains*. A domain of interest can be any subpopulation. Some domains may be very small in the sense that very few observed elements fall into it. The precision attained by an estimate made for such a domain may be unsatisfactory unless adequate measures are taken at the survey design stage.

Often in practice the domains of interest are a collection of mutually exclusive and exhaustive subpopulations. Then they form a *partition* of the population U. For example, a set of domains for a population of individuals may based on a cross-classification of sex (male, female) with eight age groups covering all ages. Then the resulting 16 domains form a partition of the whole population of individuals.

We denote the domains of interest by $U_1, \ldots, U_q, \ldots, U_Q$. Suppose that we want to estimate the total of the variable y for each domain separately. The targets of estimation are then the Q quantities $Y_1, \ldots, Y_q, \ldots, Y_Q$, where $Y_q = \sum_{U_q} y_k, q = 1, \ldots, Q$.

Define the *domain indicator* for element k by

$$\delta_{qk} = \begin{cases} 1 & \text{for} \quad k \in U_q, \\ 0 & \text{for} \quad k \in U - U_q. \end{cases} \tag{4.3}$$

The domain size is $\sum_U \delta_{qk} = N_q$. We can now express the domain total Y_q with the aid of a domain-specific study variable, y_q, whose value for element k is defined as $y_{qk} = \delta_{qk} y_k$. That is,

$$y_{qk} = \begin{cases} y_k & \text{for} \quad k \in U_q, \\ 0 & \text{for} \quad k \in U - U_q. \end{cases} \tag{4.4}$$

Then the target parameter Y_q can be expressed equivalently as the total over the whole population U of the new variable y_q, that is,

$$Y_q = \sum_U y_{qk}. \tag{4.5}$$

For estimation of Y_q, we must know the value y_{qk} for every sampled element. That is, we must be able to observe not only the value y_k but also if k is in the domain or not. We often do *not* know domain membership for every population element. An exception is when domain membership is indicated in the population frame from which we are sampling.

As mentioned in Section 2.1 we may be interested in estimating, for the population and for different domains of interest, various parameters, such as a total, a ratio of totals, a mean, and so on. More complex parameters, such as a regression coefficient or a correlation coefficient, are sometimes of interest.

Many of the more complex parameters can be expressed as a function of totals, $\psi = f(Y_1, \ldots, Y_m, \ldots, Y_M)$, where $Y_1, \ldots, Y_m, \ldots, Y_M$ are different totals and ψ is a given function. The usual principle for estimating ψ calls for each total to be replaced by its estimate, and $\hat{\psi} = f(\hat{Y}_1, \ldots, \hat{Y}_m, \ldots, \hat{Y}_M)$ to be used as the point estimator of ψ. Thus, when we know how to estimate a population total, estimating parameters that are constructed as functions of totals is straightforward. For certain types of functions ψ, software such as CLAN97 can be used for calculating the point estimates $\hat{\psi}$ and their variance estimates.

4.2.   THE HORVITZ–THOMPSON ESTIMATOR

Auxiliary information is often used to construct a good sampling design. For example, the construction of an STSI design begins with the assignment of the frame elements to a set of well-defined strata, defined, say, by sex, age groups and region, if the elements are individuals. We must thus have information about age, sex and address, allowing every element in the frame to be assigned to one and only one of the strata.

A πps design requires information in the form of a positive size measure, z_k, known for every element k in the population, as when z_k is the number of employees of enterprise k in a business survey. The design is constructed so that the inclusion probability for element k is proportional, or very nearly so, to the known value z_k, as explained by (4.2).

Once the sampling design has been fixed, the inclusion probabilities π_k and the sampling design weights $d_k = 1/\pi_k$ are fixed, known quantities. We can use them to construct the *Horvitz–Thompson estimator* (HT estimator), also called the π-*estimator*. This is given by

$$\hat{Y}_{\text{HT}} = \sum_s d_k y_k. \tag{4.6}$$

This estimator is unbiased for the total $Y = \sum_U y_k$ under any sampling design satisfying $\pi_k > 0$ for all elements k. Note that once the sampling design has been fixed, the variance and other statistical properties of $\hat{Y}_{\text{HT}}$ are also fixed. In other words, after sampling and data collection, we cannot influence the variance of $\hat{Y}_{\text{HT}}$; it is determined entirely by the choice of sampling design. Consequently, if the survey plan calls for the HT estimator, the sampling design should be chosen to meet the objective of a small variance. But this is not the only consideration. The sampling design must be practical and functional in a variety of ways. The final decision is often the result of a well-reasoned compromise.

4.3. THE GENERALIZED REGRESSION ESTIMATOR

A wider and more efficient class of estimators are those that use auxiliary information explicitly at the estimation stage. Some information may already have been used at the design stage. Denote the auxiliary vector by $\mathbf{x}^*$, and its value for element k by $\mathbf{x}_k^* = (x_{1k}, \ldots, x_{jk}, \ldots, x_{Jk})'$, a column vector with $J \geq 1$ components, constructed from one or more auxiliary variables. We assume that the population total, $\sum_U \mathbf{x}_k^*$, is accurately known. The total $\sum_U \mathbf{x}_k^*$ represents information available about population U. When the value $\mathbf{x}_k^*$ is specified in the sampling frame for every element $k \in U$, we can simply sum the values $\mathbf{x}_k^*$ to obtain $\sum_U \mathbf{x}_k^*$. (We use the notation $\mathbf{x}_k^*$ rather than the simpler $\mathbf{x}_k$ for the auxiliary vector because, from Chapter 6 onwards, $\mathbf{x}_k^*$ needs to be distinguished from a conceptually different auxiliary vector.)

Given this setting for the auxiliary information, the *theory of regression estimation* forms a basis for constructing an estimator of $Y = \sum_U y_k$. This task can be solved in different ways. An overview of the field is given in Fuller (2002).

An estimator that uses the information $\sum_U \mathbf{x}_k^*$ is the *generalized regression estimator* (GREG estimator). This estimator is explained and illustrated in several texts, for example in Chapters 6 and 7 of Särndal *et al.* (1992). It is given by

$$\hat{Y}_{\text{GREG}} = \hat{Y}_{\text{HT}} + \left(\sum_U \mathbf{x}_k^* - \sum_s d_k \mathbf{x}_k^* \right)' \mathbf{B}_{s;d} \tag{4.7}$$

where

$$\mathbf{B}_{s;d} = \left(\sum_s d_k c_k \mathbf{x}_k^* (\mathbf{x}_k^*)' \right)^{-1} \left(\sum_s d_k c_k \mathbf{x}_k^* y_k \right) \tag{4.8}$$

is a vector of regression coefficients, obtained by fitting the regression of y on $\mathbf{x}^*$, using the data $(y_k, \mathbf{x}_k^*)$ for the elements $k \in s$. In $\mathbf{B}_{s;d}$, the c_k are weights specified by the statistician. The standard choice is $c_k = 1$ for all k.

The GREG estimator is 'nearly unbiased'. The bias, although not exactly zero, is small to the point of being inconsequential, even for relatively modest sample sizes. The bias divided by the standard error tends to zero as the sample size increases.

The term $\left(\sum_U \mathbf{x}_k^* - \sum_s d_k \mathbf{x}_k^* \right)' \mathbf{B}_{s;d}$ in the formula for $\hat{Y}_{\text{GREG}}$ can be viewed as a *regression adjustment* applied to the HT estimator, $\hat{Y}_{\text{HT}} = \sum_s d_k y_k$. The effect is an important reduction of the variance of $\hat{Y}_{\text{HT}}$ when a strong regression relationship exists between y and $\mathbf{x}^*$. Then $\sum_s d_k y_k$ and $\left(\sum_U \mathbf{x}_k^* - \sum_s d_k \mathbf{x}_k^* \right)' \mathbf{B}_{s;d}$ are strongly negatively correlated, which explains the variance reduction.

We can also write $\hat{Y}_{\text{GREG}}$ as

$$\hat{Y}_{\text{GREG}} = \sum_U \hat{y}_k + \sum_s d_k (y_k - \hat{y}_k)$$

where $\hat{y}_k = (\mathbf{x}_k^*)' \mathbf{B}_{s;d}$. This form highlights the idea of prediction of the nonobserved y_k-values. Here, the prediction is obtained by linear regression fit. The point can be made that when $\mathbf{x}_k^*$ is specified in the sampling frame for every element $k \in U$, it constitutes a richer supply of information, compared to knowing 'only' $\sum_U \mathbf{x}_k^*$, which may be a value imported from a source external to the survey. Consequently, having access to $\mathbf{x}_k^*$ for every $k \in U$ gives an incentive to construct the estimator $\sum_U \hat{y}_k + \sum_s d_k (y_k - \hat{y}_k)$ with more efficient predictions $\hat{y}_k$ than $(\mathbf{x}_k^*)' \mathbf{B}_{s;d}$. One alternative is to obtain the $\hat{y}_k$ by nonparametric regression techniques, for example, local polynomial regression. Recent references are Breidt and Opsomer (2000) and Montanari and Ranalli (2003). Lehtonen and Veijanen (1998) used logistic regression fit, suitable when the y-variable is dichotomous.

Although we call $\hat{Y}_{\text{GREG}}$ as given by (4.7) and (4.8) the GREG estimator – in singular form – it is in reality a whole class of estimators, corresponding to all possible choices for the auxiliary vector $\mathbf{x}_k^*$ and the factor c_k. If a number of x-variables, each with a known population total, are available at the estimation stage, we may include in $\mathbf{x}_k^*$ those that promise to be the most efficient ones for reducing the variance. That is, we select some or all of the available x-variables for inclusion in the auxiliary vector $\mathbf{x}_k^*$. Consequently, the vector $\mathbf{x}_k^*$ to be used in $\hat{Y}_{\text{GREG}}$ can take a variety of forms. This is illustrated by several examples in Section 4.5.

Note that we can wait until after sampling and data collection to specify which of the possible GREG estimators we are going to use. The decision on the x-variables to include in $\mathbf{x}_k^*$ can be deferred until after the completion of these survey operations.

It is helpful to express the GREG estimator given by (4.7) and (4.8) as a sum of weighted observed values y_k. We have

$$\hat{Y}_{\text{GREG}} = \sum_s d_k g_k y_k \tag{4.9}$$

where the total weight given to the value y_k is the product of two weights, the design weight $d_k = 1/\pi_k$, and the weight g_k, which depends both on the element k and on the whole sample s of which k is a member. It is given by

$$g_k = 1 + \boldsymbol{\lambda}'_s c_k \mathbf{x}^*_k \tag{4.10}$$

with $\boldsymbol{\lambda}'_s = \left(\sum_U \mathbf{x}^*_k - \sum_s d_k \mathbf{x}^*_k \right)' \left(\sum_s d_k c_k \mathbf{x}^*_k (\mathbf{x}^*_k)' \right)^{-1}$. The value of g_k is near unity for a majority of the elements $k \in s$, because $\boldsymbol{\lambda}'_s$ is near the zero vector. The larger the sample size, the stronger is the tendency for the g_k to hover near unity. It is rare to find elements with a weight g_k greater than 4 or less than 0. Negative weights are in principle allowed. A few negative weights do not in any way invalidate the theory, but some users prefer all weights to be positive. Some of the existing software modifies the weights to ensure that all of them are positive.

As is easily verified, the HT estimator is a special case of $\hat{Y}_{\text{GREG}}$, obtained when (i) $\mathbf{x}^*_k = c_k = 1$ for all $k \in s$, and (ii) the design satisfies $\sum_s d_k = N$. Condition (ii) holds, for example, for the SI and STSI designs.

When we apply the weight system $d_k g_k$ for $k \in s$ to the auxiliary vector $\mathbf{x}^*_k$, and sum over s, we obtain an estimate of the population total of $\mathbf{x}^*_k$. This estimate agrees exactly with the known value of that total, that is,

$$\sum_s d_k g_k \mathbf{x}^*_k = \sum_U \mathbf{x}^*_k. \tag{4.11}$$

This property makes good sense. The weight system would not seem reasonable if it led us to estimate the total of $\mathbf{x}^*_k$ by anything other than the known value for this total. The weight system is called *calibrated* or, sometimes, *consistent*. More specifically, it is calibrated to the input of information, the population total $\sum_U \mathbf{x}^*_k$. (Some of the nonparametric regression techniques mentioned earlier in this section also give a weight system that preserves the calibration property (4.11).) From a national statistical agency's point of view, it is important to have a weight system, such as $d_k g_k$ for $k \in s$, that is applicable to all study variables in a survey.

Estimation for a domain is straightforward. We obtain the GREG estimator of the domain total $Y_q = \sum_U y_{qk}$ recognizing that the only difference lies in the study variable. The appropriate study variable for the domain is y_q given by (4.4), rather than y itself. The weights $d_k g_k$ with g_k given by (4.10) contain the information at our disposal. We obtain for Y_q the estimator

$$\hat{Y}_{q\text{GREG}} = \sum_s d_k g_k y_{qk}. \tag{4.12}$$

4.4. VARIANCE AND VARIANCE ESTIMATION

With every estimator $\hat{Y}$ is associated a variance over repeated samples. This variance, $V(\hat{Y})$, is an unknown quantity, a function of the whole population.

Since we observe only a sample, an important survey objective is to estimate the variance. The usual procedure is to start with the formula $V(\hat{Y})$ for the variance, and to transform it into an estimated variance. Once computed from the sample data, the estimated variance, denoted $\hat{V}(\hat{Y})$, provides the means for assessing the precision of $\hat{Y}$. We can use $\sqrt{\hat{V}(\hat{Y})}$ together with the point estimate $\hat{Y}$ to compute a confidence interval for the unknown parameter Y. Here and in the following sections we assume that an approximate 95 % confidence interval is computed according to the recipe

$$\text{point estimate} \pm 1.96 \, (\text{variance estimate})^{1/2}.$$

The method relies on an approximately normal distribution (over all possible samples) of the point estimator.

The variance of $\hat{Y}_{\text{HT}}$, given in many textbooks, is a quadratic form in the values y_k for $k \in U$. It is given by

$$V(\hat{Y}_{\text{HT}}) = \sum\sum_U \left(\frac{d_k d_\ell}{d_{k\ell}} - 1 \right) y_k y_\ell. \qquad (4.13)$$

(For any set of elements A, $A \subseteq U$, we write the double sum $\sum_{k \in A} \sum_{\ell \in A}$ for simplicity as $\sum\sum_A$.) The corresponding estimated variance is

$$\hat{V}(\hat{Y}_{\text{HT}}) = \sum\sum_s (d_k d_\ell - d_{k\ell}) y_k y_\ell. \qquad (4.14)$$

The GREG estimator $\hat{Y}_{\text{GREG}}$ defined by (4.9) and (4.10) has nonlinear form. Its exact variance is complex. We can linearize $\hat{Y}_{\text{GREG}}$ and find a close approximation to the variance, expressed by means of the residuals of the regression of y_k on $\mathbf{x}_k^*$. If that regression were fitted by weighted least squares using the data $(y_k, \mathbf{x}_k^*)$ for the whole population, the residuals would be

$$e_k = y_k - (\mathbf{x}_k^*)' \mathbf{B}_U \qquad (4.15)$$

where

$$\mathbf{B}_U = \left(\sum_U c_k \mathbf{x}_k^* (\mathbf{x}_k^*)' \right)^{-1} \left(\sum_U c_k \mathbf{x}_k^* y_k \right). \qquad (4.16)$$

The relation of variance to the residuals e_k becomes clear by a simple argument: we can write the deviation of $\hat{Y}_{\text{GREG}}$ from the target value Y as

$$\hat{Y}_{\text{GREG}} - Y = \sum_s d_k e_k - \sum_U e_k - R$$

where the main term is $\sum_s d_k e_k - \sum_U e_k$, with zero expected value, and $R = \left(\sum_s d_k \mathbf{x}_k^* - \sum_U \mathbf{x}_k^* \right)' (\mathbf{B}_{s;d} - \mathbf{B}_U)$ is a lower-order term, small in comparison to the main term. The expressions for $\mathbf{B}_{s;d}$ and $\mathbf{B}_U$ are (4.8) and (4.16), respectively. Dropping R, we have to close approximation

$$\hat{Y}_{\text{GREG}} - Y \approx \sum_s d_k e_k - \sum_U e_k.$$

The fact that the right-hand side has expected value zero shows that $\hat{Y}_{\text{GREG}}$ is approximately without bias for Y. Because $\sum_s d_k e_k$ is an HT estimator in the residuals e_k, we can substitute e_k for y_k in (4.13) and obtain the variance approximation

$$V(\hat{Y}_{\text{GREG}}) \approx \sum\sum_U \left(\frac{d_k d_\ell}{d_{k\ell}} - 1 \right) e_k e_\ell. \tag{4.17}$$

It is in the interest of achieving a small variance to construct the auxiliary vector $\mathbf{x}_k^*$ so as to obtain small residuals e_k. If all are near zero, the variance of $\hat{Y}_{\text{GREG}}$ should be near zero. A variance estimator is obtained by substitution in a similar manner in (4.14):

$$\hat{V}(\hat{Y}_{\text{GREG}}) = \sum\sum_s (d_k d_\ell - d_{k\ell})\hat{e}_k \hat{e}_\ell \tag{4.18}$$

where the estimated residual is

$$\hat{e}_k = y_k - (\mathbf{x}_k^*)'\mathbf{B}_{s;d} \tag{4.19}$$

with $\mathbf{B}_{s;d}$ determined by (4.8).

A sometimes used minor variation of (4.18) is

$$\hat{V}_{\text{alt}}(\hat{Y}_{\text{GREG}}) = \sum\sum_s (d_k d_\ell - d_{k\ell})(g_k \hat{e}_k)(g_\ell \hat{e}_\ell). \tag{4.20}$$

Although an argument can be made on theoretical grounds that (4.20) is more appropriate than (4.18), the two will usually differ by little, numerically speaking, because the g_k vary around unity, and (4.18) is adequate in practice. For a more detailed discussion, see, for example, Särndal *et al.* (1992). Both (4.18) and (4.20) can be computed, for important sampling designs, by software such as CLAN97.

Remark 4.1

While either (4.18) or (4.20) is satisfactory for sufficiently large samples, it is known that they have a tendency to understate the variance of $\hat{Y}_{\text{GREG}}$ when the sample is not so large. Different suggestions have been made as to how to reduce the underestimation. One avenue is to modify (4.18) or (4.20) by replacing $\hat{e}_k$ by a slightly expanded residual, $\hat{e}_{\text{adj},k} = f_k \hat{e}_k$, where the factor f_k is suitably defined. One basis for defining f_k is to make its value related to number of degrees of freedom lost in the estimation of regression parameters. Lundström (1997, Section 2.3.1), provides some suggestions.   □

Ease of computation can be an issue for variance estimation. As defined, expressions such as (4.14) and (4.18) are not conducive to quick computation because of the double sum. Both formulae have as many as $n(n + 1)/2$ different terms, if all $d_{k\ell} = 1/\pi_{k\ell}$ are unequal, as they may be for πps sampling. For example, if $n = 500$, there are around $120\,000$ terms, and to compute that many terms would be burdensome. Good approximations are available, as pointed out later in this section.

Example 4.3.   Variance estimation under STSI

For some sampling designs, (4.14) and (4.18) simplify considerably with the aid of algebraic manipulation. An example is STSI. For this design, d_k and $d_{k\ell}$ are given in Example 4.1. We have

$$\hat{Y}_{HT} = \sum_{h=1}^{H} N_h \bar{y}_{s_h}$$

with $\bar{y}_{s_h} = \sum_{s_h} y_k / n_h$. Inserting the expressions for d_k and $d_{k\ell}$ into (4.14) and simplifying, we find the factor $N_h^2 (1 - f_h)/n_h$ characteristic of stratified sampling and, finally, the well-known formula

$$\hat{V}(\hat{Y}_{HT}) = \sum_{h=1}^{H} N_h^2 \frac{1 - f_h}{n_h} S_{y s_h}^2 \qquad (4.21)$$

where $f_h = n_h / N_h$ and $S_{y s_h}^2 = \sum_{s_h} (y_k - \bar{y}_{s_h})^2 / (n_h - 1)$. The GREG estimator is

$$\hat{Y}_{GREG} = \sum_{h=1}^{H} N_h \frac{\sum_{s_h} g_k y_k}{n_h}$$

and its estimated variance, obtained from (4.18) and (4.19), is and

$$\hat{V}(\hat{Y}_{GREG}) = \sum_{h=1}^{H} N_h^2 \frac{1 - f_h}{n_h} \frac{\sum_{s_h} \left(\hat{e}_k - (1/n_h) \sum_{s_h} \hat{e}_k \right)^2}{n_h - 1}. \qquad \square$$

There are good approximations to (4.14) and (4.18) that use the d_k but dispense with the $d_{k\ell}$, so as to avoid the double sum computation. One of these is due to Deville (1993) and related to earlier derivations by Hàjek (1981). For STSI, with notation as in Example 4.1, an approximate variance estimator for $\hat{Y}_{HT}$ is

$$\hat{V}_a(\hat{Y}_{HT}) = \sum_{h=1}^{H} \frac{n_h}{n_h - 1} \sum_{s_h} (1 - \pi_k) \left(\frac{y_k}{\pi_k} - \frac{\sum_{s_h} (1 - \pi_k) y_k / \pi_k}{\sum_{s_h} (1 - \pi_k)} \right)^2 \qquad (4.22)$$

where π_k is the inclusion probability of element k within stratum h. When πps sampling is applied in stratum h, we have $\pi_k = n_h z_k / \sum_{U_h} z_k$, where z_k is the size measure and n_h is the (expected) stratum sample size. Variance estimation in software such as POULPE and CLAN97 is executed in close agreement with the approximation (4.22).

Finally, we need to examine variance estimation for the domain estimator $\hat{Y}_{qGREG}$ given by (4.12). A simple modification of (4.18) gives the variance estimator: we simply replace y_k by y_{qk}, given by (4.4). Consequently, $\hat{e}_k$ in (4.18) is replaced by the modified residual

$$\hat{e}_{qk} = y_{qk} - (\mathbf{x}_k^*)' \mathbf{B}_{(q)s;d} \qquad (4.23)$$

where $\mathbf{B}_{(q)s;d}$ is the expression obtained from (4.8) when we replace y_k by y_{qk}. The basis for the procedure is that the GREG estimation technique (which includes point estimation and variance estimation) is applicable to any study variable having constant numbers $y_1, \ldots, y_N$ attached to the N population elements. It works, in particular, for the study variable y_q whose values are defined by (4.4), despite its unusual character of being a zero-valued variable for many elements. From (4.23) it can be seen why it is important to identify an auxiliary vector that comes close to identifying the domains. If it did not, many residuals (4.23) could be large, and the benefit of the auxiliary information might be slight, especially for small domains. Example 4.6 illustrates this issue.

4.5. EXAMPLES OF THE GENERALIZED REGRESSION ESTIMATOR

The examples in this section illustrate the GREG estimator given by (4.9) and (4.10) in selected special cases, corresponding to simple auxiliary vectors.

Example 4.4. One-way classification

For a population of individuals, suppose we know the frequency of men and women, N_1 and N_2. The appropriate formulation of the auxiliary vector $\mathbf{x}_k^*$ recognizes the two categories: $\mathbf{x}_k^* = (1, 0)'$ for all men, and $\mathbf{x}_k^* = (0, 1)'$ for all women. The population total of $\mathbf{x}_k^*$ is thus $(N_1, N_2)'$. From (4.10) we obtain $g_k = N_1 / \sum_{s_1} d_k$ for every man and $g_k = N_2 / \sum_{s_2} d_k$ for every woman, where s_1 and s_2 denote the male part and the female part, respectively, of the sample s. As is easily verified, the weights $d_k g_k$ satisfy the calibration requirement $\sum_s d_k g_k \mathbf{x}_k^* = (N_1, N_2)'$. The GREG estimator becomes $\hat{Y}_{\text{GREG}} = N_1 \bar{y}_{s_1;d} + N_2 \bar{y}_{s_2;d}$ where the d-weighted means are $\bar{y}_{s_j;d} = \sum_{s_j} d_k y_k / \sum_{s_j} d_k$ for $j = 1, 2$. The form that GREG takes in this case is known as the *poststratified estimator*. Here there are two poststrata, men and women. The generalization to an arbitrary number of poststrata is straightforward. ☐

In the next example, the categories of the auxiliary vector are defined by a cross-classification.

Example 4.5. Two-way classification

Consider a population of individuals listed in a register that specifies, for every person, sex and one of three possible regions. Then all population counts in the following table can be derived, the six cell counts and the five marginal counts:

		Region			
Sex		1	2	3	Total
Male	1	N_{11}	N_{12}	N_{13}	$N_{1\bullet}$
Female	2	N_{21}	N_{22}	N_{23}	$N_{2\bullet}$
Total		$N_{\bullet 1}$	$N_{\bullet 2}$	$N_{\bullet 3}$	N

In its most detailed form, the auxiliary information consists of the six cell counts $N_{11}, \ldots, N_{23}$, and the associated auxiliary vector has dimension 6. All entries are zero except the one which, with a value of 1, identifies the cell to which k belongs. For example, for every person in the cell 'males in Region 2', $\mathbf{x}_k^* = (0, 1, 0, 0, 0, 0)'$. The population sum of these N vectors is the known vector $(N_{11}, N_{12}, N_{13}, N_{21}, N_{22}, N_{23})'$. The GREG estimator $\hat{Y}_{\mathrm{GREG}}$ obtained by this $\mathbf{x}_k^*$-vector is a poststratified estimator with six terms, corresponding to six poststrata.

But we may choose not to cross-classify. Situations occur where complete cross-classification of two or more categorical variables is impractical or inconvenient, for instance, when (i) the variables come from different registers, or (ii) some cell counts are small. In situation (i), a decision to use cell counts may incur a cost, because registers may have to be matched to establish that information. In situation (ii), small cell counts may cause the estimator to be unstable. This can sometimes be avoided by collapsing cells.

An option is to use only the information defined by the marginal counts. The auxiliary vector for this case is of dimension 5. Its first two positions are used to code 'sex', and the final three are used to code 'region'. For example, the auxiliary vector for every person in the cell 'males in Region 2' is $\mathbf{x}_k^* = (1, 0, 0, 1, 0)$. The required population sum of the $\mathbf{x}_k^*$-vectors is $(N_{1\bullet}, N_{2\bullet}, N_{\bullet 1}, N_{\bullet 2}, N_{\bullet 3})'$, the vector of known marginal counts. We have occasion to further examine the two-way classification in Section 7.6. □

A principle in estimation for domains is to use an auxiliary vector that identifies, as closely as possible, the domains. This will tend to reduce the magnitude of the residuals, thus reducing the variance. The following example illustrates this.

Example 4.6. Domain estimation

For a population of individuals, suppose that the survey requires separate estimates to be made for males and females. They define two domains of the population, U_q for $q = 1, 2$. Further, let us assume SI with n elements drawn from N, and that we know the number of men and women in the population. We have decided to use the GREG estimator for the domain total $Y_q = \sum_{U_q} y_k = \sum_U y_{qk}, q = 1, 2$. We must then specify the auxiliary vector. Let us consider two alternatives:

(i) The simplest possible specification, $\mathbf{x}_k^* = c_k = 1$ for all elements. Then $\hat{Y}_{q\mathrm{GREG}} = (N/n) \sum_{s_q} y_k$, where $s_q = s \cap U_q$ is the part of the sample s that happens to fall in the domain U_q.
(ii) Let $\mathbf{x}_k^* = (1, 0)'$ for all males, $\mathbf{x}_k^* = (0, 1)'$ for all females, and $c_k = 1$ for all k. Then $\hat{Y}_{q\mathrm{GREG}} = N_q \bar{y}_{s_q} = N_q \left(\sum_{s_q} y_k \right) / n_q$, where N_q and n_q are the respective sizes of U_q and s_q.

For both alternatives we obtain an approximate variance of the form

$$V(\hat{Y}_{q\mathrm{GREG}}) \approx N^2 \frac{1 - n/N}{n} \frac{1}{N - 1} \sum_U e_{qk}^2.$$

The only difference between the two alternatives lies in the residuals e_{qk}. In alternative (i) they are

$$e_{qk} = \begin{cases} y_k - Y_q/N & \text{for } k \in U_q, \\ -Y_q/N & \text{for } k \in U - U_q; \end{cases}$$

and in alternative (ii) they are

$$e_{qk} = \begin{cases} y_k - Y_q/N_q & \text{for } k \in U_q, \\ 0 & \text{for } k \in U - U_q. \end{cases}$$

It is easily seen that $\sum_U e_{qk}^2$ (and therefore the variance) is considerably greater for (i) than for (ii). The reason is that (i) forgoes some information that is actually available, namely, the known sizes $N_q, q = 1, 2$.

In alternative (ii), the auxiliary vector coincides exactly with the domain indicator. It identifies the domains without any discrepancies, whereas the vector in (i) does not recognize the domains at all. Thus (ii) achieves a significant variance reduction compared to (i). □

The examples given so far in this section involve categorical auxiliary information. The next example combines a quantitative auxiliary variable x_k with a categorical variable. A number of possibilities arise for formulating the auxiliary vector $\mathbf{x}_k^*$. The variable x_k can be used by itself or in different combinations with the categorical variables.

Example 4.7. A one-way classification combined with a quantitative variable

Assume that the frame specifies sex and region for every person k, as in Example 4.5, and, in addition, the value x_k of a quantitative auxiliary variable, such as the income of person k. The six population cells are denoted $U_{11}, U_{12}, \ldots, U_{23}$. The three regions are $U_{\bullet 1}, U_{\bullet 2}$ and $U_{\bullet 3}$. We can compute six cell counts and the corresponding six cell totals of x_k. Depending on whether we use all of this information or only part of it, we obtain a number of possible formulations of the auxiliary vector $\mathbf{x}_k^*$. Five alternatives are shown in the following table:

Case	Auxiliary vector $\mathbf{x}_k^*$	Auxiliary population total $\sum_U \mathbf{x}_k^*$
(i)	x_k	$\sum_U x_k$
(ii)	$(1, x_k)'$	$\left(N, \sum_U x_k\right)'$
(iii)	$(0, x_k, 0, 0, 0, 0)'$	$\left(\sum_{U_{11}} x_k, \ldots, \sum_{U_{23}} x_k\right)'$
(iv)	$(0, 1, 0, 0, 0, 0, 0, x_k, 0, 0, 0, 0)'$	$\left(N_{11}, \ldots, N_{23}, \sum_{U_{11}} x_k, \ldots, \sum_{U_{23}} x_k\right)'$
(v)	$(1, 0, 0, x_k, 0)'$	$\left(N_{1\bullet}, N_{2\bullet}, \sum_{U_{\bullet 1}} x_k, \sum_{U_{\bullet 2}} x_k, \sum_{U_{\bullet 3}} x_k\right)'$

In case (iv), in the vector $\mathbf{x}_k^*$ of dimension 12, cell membership is indicated by the first six positions, and the final six positions are used to place the x_k-value in the corresponding cell position. The particular form of $\mathbf{x}_k^*$ shown in the table is for 'males in region 2'. It is case (iv) that makes the most extensive or complete use of the available information, the six cell counts and the six cell x_k -totals. We can expect (iv) to produce the lowest variance for the five GREG estimators.

In case (v), the vector $\mathbf{x}_k^*$ of dimension 5 uses the first two positions to code sex, and the final three put the x_k -value in the region where k belongs. For 'males in region 2' we get the appearance of $\mathbf{x}_k^*$ shown in the table. Here, the information used consists of the population counts for men and women, and the x_k -totals by region. This falls short of a complete use of the available information.

We wish to emphasize that a given quantity of auxiliary information may lead to several different formulations of the auxiliary vector $\mathbf{x}_k^*$. Some well-known estimator types are among those generated by the five formulations listed. Let us consider a few of them.

When $\mathbf{x}_k^* = x_k$ and $c_k = 1/x_k$, the general GREG formula (4.9) gives the *ratio estimator*

$$\hat{Y}_{\text{GREG}} = N\overline{x}_U \frac{\overline{y}_{s;d}}{\overline{x}_{s;d}} \qquad (4.24)$$

where $\overline{x}_U = \sum_U x_k/N, \overline{y}_{s;d} = \sum_s d_k y_k / \sum_s d_k$ and $\overline{x}_{s;d}$ is the analogously d-weighted sample mean for the x-variable. (In the notation the first index shows the set over which the quantity is computed, and the second index shows the weighting. Hence, $\overline{y}_{s;d}$ is computed over the sample s, with design weighting as defined by the d_k. Quantities obtained by equal weighting, such as $\overline{x}_U$, are given only the first index.)

When $\mathbf{x}_k = (1, x_k)'$ and $c_k = 1$ for all k, (4.9) gives another well-known form, the (simple) *regression estimator*

$$\hat{Y}_{\text{GREG}} = N\{\overline{y}_{s;d} + (\overline{x}_U - \overline{x}_{s;d})B_{s;d}\} \qquad (4.25)$$

with $B_{s;d} = \sum_s d_k(x_k - \overline{x}_{s;d})(y_k - \overline{y}_{s;d}) / \sum_s d_k(x_k - \overline{x}_{s;d})^2$. $\qquad \square$

CHAPTER 5

Introduction to Estimation in the Presence of Nonresponse

5.1. GENERAL BACKGROUND

As in Chapter 4, we seek to estimate the target population total of the study variable y, $Y = \sum_U y_k$, and the totals of specified domains, $Y_q = \sum_{U_q} y_k$ for $q = 1, \ldots, Q$. To this end, a sample s is drawn from the frame according to a given sampling design. As in Chapter 4, we assume in this chapter that the frame population agrees exactly with the target population, U. In practice this condition is often not met, because of frame errors. These imperfections are discussed in Chapter 14. A difference in this chapter compared with Chapter 4 is that we no longer assume complete response.

The given sampling design determines the design weight $d_k = 1/\pi_k$ for every element $k \in U$, where π_k is the inclusion probability of k. Most surveys involve more than one study variable, perhaps many. We can think of each study variable as an item on a questionnaire distributed to the elements k of the sample s. The response set for the survey is denoted r. As mentioned in Example 2.3, it is the set of all elements for which we have, at the end of the data collection, an observed response for at least one of the questionnaire items. The unit nonresponse is the complement set, $s - r$. The situation is illustrated in Figure 5.1.

For a particular variable (or questionnaire item) y, the set of elements for which y_k is missing defines the nonresponse set for that particular variable. This set contains all of the unit nonresponse, $s - r$, and an additional set of elements, namely, those with item nonresponse on variable y.

If the survey has item nonresponse as well as unit nonresponse, important decisions must be made about how item nonresponse is to be treated in the estimation. Without loss of generality, we can defer this discussion until Chapters 12 and 13.

The literature proposes two main approaches for dealing with nonresponse, *weighting* and *imputation*. When weighting is used, a set of weights is determined with the aid of the available auxiliary information, and estimation is carried out by applying the weights to the y-values for the responding elements. In this book, we use the *calibration approach* to compute the weights.

Estimation in Surveys with Nonresponse   C.-E. Särndal and S. Lundström
© 2005 John Wiley & Sons, Ltd

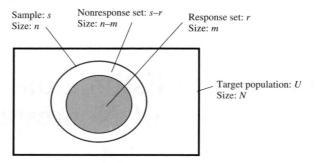

Figure 5.1   Representation of selected sample and response set, seen as subsets of the target population, which is assumed identical to the frame population.

If y is affected by unit nonresponse alone, and no item nonresponse, the estimator of the parameter of interest, $Y = \sum_U y_k$, will be $\hat{Y}_W = \sum_r w_k y_k$, where the index W is for 'weighting' and the calibrated weights w_k are, at least for most elements, greater than the weights that would be appropriate for complete response. This is to compensate for elements lost due to nonresponse. Chapters 6–11 provide a systematic account of weighting using the calibration approach.

The other principal approach for nonresponse treatment, imputation, implies that proxy values are created for the values y_k that are missing because of nonresponse. The proxy value for element k, the *imputed value* for k, is denoted $\hat{y}_k$. The superimposed 'hat' serves as a reminder that an imputed value is in some sense an estimated value, rather than one that has been properly observed. For element k, the 'estimation' is carried out by a given *imputation method*. There are many imputation methods in current use, and more than one imputation method is often used in one and the same survey. In other words, not all $\hat{y}_k$ may be constructed by the same method. Chapter 12 contains a detailed discussion of different imputation procedures.

We shall pay particular attention to a combination of imputation and weighting such that imputation is used for the item nonresponse, while a calibrated weighting compensates for the unit nonresponse. To this end, we first create a completed data set consisting of values, some observed, some imputed, for every survey variable and every element k in the survey response set r. For a particular variable y, the completed data set can be written as $\{y_{\bullet k} : k \in r\}$, where $y_{\bullet k} = y_k$, an observed value, if element k responds to the variable y, and $y_{\bullet k} = \hat{y}_k$, an imputed value, if y_k is missing through item nonresponse and imputed by $\hat{y}_k$.

The estimator of $Y = \sum_U y_k$ is then computed from the completed data set as $\hat{Y}_{IW} = \sum_r w_k y_{\bullet k}$, where the calibrated weights w_k compensate for the unit nonresponse. The double index IW indicates 'imputation followed by weighting'.

5.2. ERRORS CAUSED BY SAMPLING AND NONRESPONSE

We shall consider techniques for simultaneously reducing the sampling error and the nonresponse error *after the data collection stage*, that is, after some nonresponse has occurred. These techniques rely on an effective use of powerful auxiliary information. Our general notation will be $\hat{Y}_W$ for a calibration estimator and $\hat{Y}_{IW}$ for an estimator created by imputation followed by calibrated weighting.

Within each approach, there are many possible uses of auxiliary information. Even with the use of powerful auxiliary information, we have to accept that both $\hat{Y}_W$ and $\hat{Y}_{IW}$ are affected by sampling error and nonresponse error. However, generally speaking, the better the auxiliary information, the smaller the two errors.

The two types of error were briefly mentioned in Section 2.2. Our objective in this section is to look more closely at these errors and the procedures used to reduce them.

In this section we let the notation $\hat{Y}_{NR}$, for *nonresponse estimator*, represent either $\hat{Y}_W$ (if there is weighting only) or $\hat{Y}_{IW}$ (if there is imputation followed by weighting). Further, let us denote by $\hat{Y}$ the expression taken by $\hat{Y}_{NR}$ for the case of complete response, that is, when $r = s$. In the methods we consider, $\hat{Y}$ is a GREG estimator, an HT estimator or some other estimator that is unbiased, or very nearly so, under repeated samples s drawn from U.

The error of $\hat{Y}_{NR}$ is its deviation from the target parameter value $Y = \sum_U y_k$. We can write the error as a sum of two components,

$$\hat{Y}_{NR} - Y = (\hat{Y} - Y) + (\hat{Y}_{NR} - \hat{Y}). \tag{5.1}$$

The first term on the right hand side, $\hat{Y} - Y$, is the *sampling error* (the error caused by selecting and observing a sample only, as opposed to the whole population). The second term, $\hat{Y}_{NR} - \hat{Y}$, is the *nonresponse error* (the error incurred because complete response has not been achieved). A survey usually has other nonnegligible errors, but they are not considered here.

We need to know the basic statistical properties of the estimator $\hat{Y}_{NR}$. The *central tendency* of $\hat{Y}_{NR}$ is determined by its expected value. The *accuracy* of $\hat{Y}_{NR}$ is determined by its mean squared error. To derive these quantities we must average over the possible outcomes.

In the present setting, the notions of expected value, unbiasedness, variance and mean squared error are tied to a twofold averaging process: over all possible samples s that can be drawn by the known sampling design, denoted $p(s)$; and over all possible response sets r that can be realized, given s, under the unknown response distribution, denoted $q(r|s)$.

We denote the expectation operators with respect to these two distributions by E_p and E_q respectively. Operators with respect to both distributions jointly will be given the index pq.

The average (over all possible samples s) of the sampling error $\hat{Y} - Y$ is zero or nearly zero. The average (over all possible samples s *and* all possible response sets r) of the nonresponse error $\hat{Y}_{NR} - \hat{Y}$ is likely to differ from zero, because nonresponse invariably causes some bias.

The conditional (on s) expected nonresponse error, which we can call the *nonresponse bias given the sample*, is

$$B_{NR|s} = E_q\left((\hat{Y}_{NR} - \hat{Y})|s\right) = E_q(\hat{Y}_{NR}|s) - \hat{Y}. \tag{5.2}$$

Averaging over all possible samples s, we obtain the (unconditional) *nonresponse bias* of $\hat{Y}_{NR}$,

$$B_{NR} = E_p(B_{NR|s}) = E_{pq}(\hat{Y}_{NR} - \hat{Y}) = E_{pq}(\hat{Y}_{NR}) - E_p(\hat{Y}).$$

The expected sampling error, which we can call the *sampling bias*, is

$$B_{SAM} = E_p(\hat{Y} - Y) = E_p(\hat{Y}) - Y.$$

The (unconditional) bias of $\hat{Y}_{NR}$, is

$$B_{pq}(\hat{Y}_{NR}) = E_{pq}(\hat{Y}_{NR} - Y) = B_{SAM} + B_{NR}. \tag{5.3}$$

The bias of the estimator $\hat{Y}_{NR}$ is the sum of two components, the p-expectation of the sampling error (the sampling bias) and the pq-expectation of the nonresponse error (the nonresponse bias). The first of these is zero or negligible for most practical purposes. The bias of $\hat{Y}_{NR}$ comes entirely or almost entirely from the unknown (and perhaps significant) nonresponse bias.

The bias $B_{pq}(\hat{Y}_{NR})$ is at the centre of attention in this book. In practice, it is virtually impossible to tell its magnitude, because the response distribution $q(r|s)$ is never exactly known. However, in practice one frequently assumes, rightly or wrongly, that the nonresponse bias is 'sufficiently small'. More often than not, the assumption is unjustified, and confidence interval statements, for example, will be invalid.

Much of our insight into the nonresponse bias must come from simulation studies where the bias of specific estimators is evaluated for different populations and under different assumed response distributions. Such simulation studies are illustrated in Section 7.7.

The bias (5.3) cannot be estimated, at least not satisfactorily. We derive in Chapter 9 an expression for the bias of the calibration estimator $\hat{Y}_W$. It shows the bias as a function of the values y_k, $\mathbf{x}_k$ and θ_k of all $k \in U$. It is a theoretical, nonestimable expression. Nevertheless, it can be a valuable guide in the search for the best possible auxiliary vector $\mathbf{x}_k$, with a view to control the bias of $\hat{Y}_W$. Another type of analysis, illustrated in Chapter 10, consists in examining how well specific estimators $\hat{Y}_W$, corresponding to particular auxiliary vectors $\mathbf{x}_k$, can succeed in protecting against nonresponse bias.

We turn to the variance, $V_{pq}(\hat{Y}_{NR}) = E_{pq}\left(\hat{Y}_{NR} - E_{pq}(\hat{Y}_{NR})\right)^2$. The variance can be written in easily understood form as sum of two components,

$$\text{Variance} = \text{Sampling variance} + \text{Nonresponse variance}.$$

More specifically, if the conditional bias $B_{NR|s}$ is zero for every possible s, then

$$V_{pq}(\hat{Y}_{NR}) = V_{SAM} + V_{NR} \tag{5.4}$$

where $V_{SAM} = V_p(\hat{Y})$ is called the *sampling variance*, and $V_{NR} = E_p V_q(\hat{Y}_{NR}|s)$ is called the *nonresponse variance*.

The result (5.4) is shown in the end-of-chapter appendix. The component V_{SAM} is the variance of $\hat{Y}$ over all possible samples s that can be drawn with the given sampling design; it does not depend on the nonresponse or on the response distribution. The component V_{NR} involves averaging over all samples s as well as over all response sets r.

The requirement $B_{NR|s} = 0$ for every possible s is a strong one. If it does not hold, further components are added to the nonresponse variance, so that V_{NR} in (5.4) becomes

$$V_{NR} = E_p V_q(\hat{Y}_{NR}|s) + V_p(B_{NR|s}) + 2\text{Cov}_p(\hat{Y}, B_{NR|s}). \tag{5.5}$$

Equation (5.4) states that even in the fortunate circumstance that nonresponse causes no bias, the effect will be an increase in the variance, compared to complete response, in which case we would have only the component $V_{SAM} = V_p(\hat{Y})$. Furthermore, if there is bias, the variance suffers further inflation, through the last two terms on the right-hand side of (5.5).

In practice, what we can hope for is that the estimation procedure will have succeeded in reducing the bias to low levels, so that $B_{NR|s} = 0$ holds at least approximately for every sample s.

In order to assess the probable error of $\hat{Y}_{NR}$, we need to estimate the variance. Chapter 11 suggests a procedure for the calibration estimator $\hat{Y}_W$. It is not built on a term-by-term estimation of (5.4), but follows a different line of argument. Chapter 13 discusses variance estimation for the item imputed estimator $\hat{Y}_{IW}$.

It is of obvious interest for practice to assess the nonresponse variance V_{NR}. A routine, well-founded measurement of this variance component is not standard in surveys. For example, if it were found in a regularly repeated survey that V_{NR} accounts for a significant share of the total variance, it is a clear signal to allocate further resources to reducing the nonresponse at the next survey occasion.

APPENDIX: VARIANCE AND MEAN SQUARED ERROR UNDER NONRESPONSE

The variance expression (5.4) for $\hat{Y}_{NR}$ is derived as follows. There are two phases of selection: s is selected from U, and, given s, r is realized as a subset from s. The two probability distributions are $p(s)$ and $q(r|s)$. Under these conditions, we can obtain the variance of a random quantity by the rule 'the variance of the conditional expectation plus the expectation of the conditional variance'. By (5.2), the conditional expectation is $E_q(\hat{Y}_{NR}|s) = \hat{Y} + B_{NR|s}$. We obtain the pq-variance

$$V_{pq}(\hat{Y}_{NR}) = V_p(\hat{Y} + B_{NR|s}) + E_p V_q(\hat{Y}_{NR}|s). \tag{A5.1}$$

If $B_{NR|s} = 0$ for every s, we obtain the result (5.4) with $V_{NR} = E_p V_q(\hat{Y}_{NR}|s)$. If this condition does not hold, then we expand the term $V_p(\hat{Y} + B_{NR|s})$ in (A5.1) to find the result (5.4) with V_{NR} given by (5.5).

In the presence of bias, a more relevant indicator of variability than the variance is the mean squared error (MSE). The pq-MSE of $\hat{Y}_{NR}$, $MSE_{pq}(\hat{Y}_{NR})$, is the average of the squared error, $(\hat{Y}_{NR} - Y)^2$, over all samples s and all response sets r contained in s. Using the fact that $MSE_{pq}(\hat{Y}_{NR}) = V_{pq}(\hat{Y}_{NR}) + \left(B_{pq}(\hat{Y}_{NR})\right)^2$, we obtain, after simplification,

$$MSE_{pq}(\hat{Y}_{NR}) = V_p(\hat{Y}) + E_p V_q(\hat{Y}_{NR}|s) + E_p(B^2_{NR|s}) + 2\text{Cov}_p(\hat{Y}, B_{NR|s})$$

$$+ 2B_{SAM}B_{NR} + (B_{SAM})^2. \qquad (A5.2)$$

Here, $V_p(\hat{Y})$ is the variance of the complete response estimator. The principal terms on the right-hand side are the first three. Thus we can write:

$$MSE_{pq}(\hat{Y}_{NR}) \approx V_p(\hat{Y}) + E_p V_q(\hat{Y}_{NR}|s) + E_p(B^2_{NR|s}).$$

Here, the term $E_p(B^2_{NR|s})$, caused by nonresponse bias, may represent a considerable addition to the MSE.

CHAPTER 6

Weighting of Data in the Presence of Nonresponse

6.1. TRADITIONAL APPROACHES TO WEIGHTING

Design-based theory, also called randomization theory, is applicable when the survey has complete response. This theory of inference relies on the randomization built into the sample selection. The sampling design can be stratified simple random sampling, stratified cluster sampling, and so on. Whatever it is, every element possesses its own individual, known inclusion probability π_k. These probabilities play a decisive role in statistical inference for complete response.

When nonresponse also enters the picture, as is unavoidable in practice, it is convenient to think of each element as having its own individual probability of responding. Unlike the sampling phase, the response phase is beyond the control of the statistician. Nonresponse occurs with unknown probabilities. Nevertheless, the concept of individual response probabilities creates a fruitful framework for nonresponse treatment methods, as this and following chapters attempt to show.

Our principal concern in the rest of the book is with surveys affected by nonresponse. That is, the values of the study variable y are observed only for the elements k in a subset r of the full sample s. We call r the *response set*. At this point we assume that y is affected by unit nonresponse only. Whatever the technique used, an estimator of the total $Y = \sum_U y_k$ will be more or less biased (unless the nonresponse occurs completely at random, an unlikely occurrence). Some bias is unavoidable. We seek methods to limit the bias as far as possible.

Prior to around 1980, nonresponse adjustment methods typically relied on a *deterministic model* of survey response. The finite population was assumed to consist of two nonoverlapping parts, a response stratum and a nonresponse stratum. Every element in the former was assumed to respond with certainty if selected for the sample, and every element in the latter stratum had probability zero of responding. An obvious criticism of that model is that it is simplistic and unrealistic. Moreover, the sizes of the two strata could usually not be assumed to be known, so an approach sometimes used was to estimate the total for the response stratum, and to add a term that would somehow compensate for the contribution of the nonresponse stratum.

Estimation in Surveys with Nonresponse   C.-E. Särndal and S. Lundström
© 2005 John Wiley & Sons, Ltd

In the 1980s, a more satisfactory approach became popular, which views the response set r as the result of two probabilistic selections. The sample s is first selected from the population U, then the response set r is realized as a subset of s. The approach is more realistic and more general than the deterministic one in that it allows every element k to have its own individual response probability θ_k where $0 \leq \theta_k \leq 1$ for all k. This generality is not without a price: the response probabilities θ_k are always unknown. An approach that has been used is to replace the θ_k with estimates based on auxiliary information.

Response or not on the part of element k is a binary outcome with unknown probability, just as a toss with a bent coin can give heads or tails with unknown probability. For element k, we can define a binary random variable R_k with value $R_k = 1$ if k responds and value $R_k = 0$ if not. The expected value of the response indicator R_k, conditionally on the realized sample s, is $E_q(R_k|s) = \theta_k$, where E_q is the expectation operator with respect to the response mechanism $q(r|s)$. We shall assume that the response probability θ_k depends on k but not on the sample s of which k is a member.

It is not clear who originated the important notion that we can associate with every element a unique number, the probability of responding (or of delivering the solicited value y_k of the study variable y). Early uses of this concept for analysing the properties, such as expected value and variance, of traditional estimators are found in Lindström and Lundström (1974) and Platek et al. (1978).

We can consider an estimation theory built around the idea that the element k is equipped with a known individual inclusion probability, π_k, and an unknown individual response probability, θ_k. This becomes a 'quasi-randomization theory', a fitting term coined by Oh and Scheuren (1983).

Important work was accomplished around 1980 by the Panel on Incomplete Data, sponsored by the US National Science Foundation. This resulted in three volumes of contributions by different authors, some theoretical, some empirical and practical. Several contributions are in the spirit of the quasi-randomization theory, for example, the chapter by Platek and Gray (1983). This theory has served as a guiding light in recent decades for methodologists in statistical agencies.

Other contributions in the three volumes make fruitful use of concepts, rather novel at the time, such as ignorable, as opposed to non-ignorable, nonresponse (or missingness), missing at random (MAR), missing completely at random (MCAR) are used and explained. These concepts have since contributed significantly to the understanding of nonresponse and missing data. They were formulated originally in a Bayesian framework and are discussed in sources such as Rubin (1987) and Little and Rubin (1987). In Section 9.6 we discuss these concepts and their role for this book.

The presentation in Chapters 6–14 of this book considerably extends the material in Lundström and Särndal (2001), a Current Best Methods manual produced for and used at Statistics Sweden in the treatment of survey data affected by nonresponse and frame deficiencies.

The quasi-randomization idea leads to the *two-phase approach to weighting for nonresponse*. This term alludes to the idea of selection of a sample s followed by the realization of a response set r as a subset of s. The two-phase approach for nonresponse adjustment is discussed in the recent literature, for example, in Särndal *et al.* (1992). Chapter 9 of that book deals with the traditional formulation of two-phase sampling, in the absence of nonresponse. A first sample is selected from U, certain useful variables (although not the study variable(s)), are observed, then a smaller subsample is taken from the first sample, and the study variable(s) are observed for the elements of the subsample. In this context, all inclusion probabilities are known by design, those for the first phase as well as those for the second phase. In Chapter 15 of the book, the two-phase sampling theory is adapted to the situation of main concern here: sampling is followed by nonresponse, viewed as a second phase of selection, conditional on the probability sample taken, and the obstacle of unknown response probabilities is resolved in a well-reasoned manner.

We briefly comment on the two-phase approach. It begins with the assumption that the response distribution $q(r|s)$ is known (although in practice this is not the case). This implies that the first- and second-order response probabilities,

$$\Pr(k \in r|s) = \theta_k, \qquad \Pr(k \ \& \ \ell \in r|s) = \theta_{k\ell} \qquad (6.1)$$

are known and usable in estimating $Y = \sum_U y_k$. We could then compute the combined weights $d_k/\theta_k = (1/\pi_k) \times (1/\theta_k)$ for $k \in r$ and use them to form the unbiased two-phase estimator

$$\hat{Y} = \sum_r (d_k/\theta_k) y_k. \qquad (6.2)$$

This extends the basic Horvitz–Thompson estimator to a selection in two phases. But the θ_k are always unknown in practice, so (6.2) cannot be computed. To make it operational we must take the preliminary step of estimating the unknown θ_k. Suppose auxiliary information exists to make this step possible. Let $\hat{\theta}_k$ be our estimate of θ_k, for $k \in r$. We obtain a two-phase nonresponse adjusted estimator by replacing θ_k by $\hat{\theta}_k$ in (6.2). We can then compute

$$\hat{Y} = \sum_r (d_k/\hat{\theta}_k) y_k. \qquad (6.3)$$

Different ways of deriving the estimates $\hat{\theta}_k$ have been proposed. The approach has had a considerable appeal, and estimators of the type (6.3) have been extensively studied and used in the last 30 years.

A more recent trend emphasizes the use of auxiliary information to meet two objectives: a reduction of the variance of an estimator and a reduction of its bias due to nonresponse. One can distinguish two types of auxiliary information: vectors with information up to the level of the population, so that the total of a vector $\mathbf{x}_k^*$ is known; and vectors with information extending only up to the level of the sample, so that a vector value $\mathbf{x}_k^\circ$ is specified for every $k \in s$. The use of those two types of information is explored, via regression fitting, in Särndal *et al.* (1992).

We can extend the use of the GREG estimation technique in Section 4.3. Let $\sum_U \mathbf{x}_k^*$ denote the known total of the auxiliary vector $\mathbf{x}_k^*$. If the θ_k were known, we could form the nearly unbiased two-phase GREG estimator given by

$$\hat{Y} = \sum_r \frac{d_k}{\theta_k} g_{k\theta} y_k \tag{6.4}$$

where $d_k = 1/\pi_k$ and

$$g_{k\theta} = 1 + \left(\sum_U \mathbf{x}_k^* - \sum_r \frac{d_k}{\theta_k} \mathbf{x}_k^* \right)' \left(\sum_r \frac{d_k}{\theta_k} c_k \mathbf{x}_k^* (\mathbf{x}_k^*)' \right)^{-1} (c_k \mathbf{x}_k^*) \tag{6.5}$$

where the c_k are specified constants. Replacing the unknown θ_k by suitable estimates $\hat{\theta}_k$, based on auxiliary variable values known for $k \in s$, we obtain from (6.4)

$$\hat{Y} = \sum_r \frac{d_k}{\hat{\theta}_k} g_{k\hat{\theta}} y_k \tag{6.6}$$

where $g_{k\hat{\theta}}$ is given by (6.5) with θ_k replaced by $\hat{\theta}_k$.

In general, (6.6) is preferred over (6.3), and all the more so if $\mathbf{x}_k^*$ is a powerful auxiliary vector. Special cases of (6.3) and (6.6) have been used in surveys, sometimes with good results. In this book we need not focus on (6.3) and (6.6). Many of the estimators derivable from these formulae are special cases of the wider family of calibration estimators to be considered.

Kott (1994) discusses several issues in relation to (6.2) and (6.4). These formulae contain the θ_k. They are not estimators for use in practice, because the θ_k are unknown. Also, to be unbiased or nearly so, both require that all θ_k are strictly positive. As Kott (1994) notes, this requirement is perhaps unrealistic, because most surveys have a fraction of hard core nonresponse, composed of elements that do not under any circumstances respond.

A popular approach has nevertheless been to take the impracticable formulae (6.2) and (6.4) as a point of departure and arrive at the estimators (6.3) and (6.6) through an estimation of the unknown θ_k. As a result, $d_k/\hat{\theta}_k$ will play the role of a weight for the responding element k. This is a structured approach in three steps: (i) we assume the existence of a *response mechanism*; (ii) we formulate for this mechanism a realistic model in which the unknown θ_k appear as unknown parameters; and (iii) we estimate the θ_k, using any relevant auxiliary variables and the fact that a subset of the sample s was observed to respond, whereas the complement set did not. A criticism that can be levelled against this approach is that it is hard to defend any proposed model as being more realistic than any competitor. Our knowledge about true response behaviour is limited. The time and the resources needed for in-depth study of the response behaviour are limited in statistical agencies.

Different models, some simple, some more advanced, have been used for the response mechanism. One frequently used model states that the population consists of nonoverlapping groups such that all elements within one and the same group respond with the same probability, and in an independent manner. Such groups are known as *response homogeneity groups* (RHGs). The auxiliary information required here is that we can uniquely classify every sampled element, respondent or nonrespondent, into one of the groups. Often used in surveys of individuals are groups formed by a crossing of age categories with sex, so that, for example, five age categories crossed with two sexes gives ten groups. Experience has shown that such a simple grouping will seldom accomplish the inherent assumption of constant response probability within each group. An age by sex grouping may be easy to establish but is usually inefficient for explaining a response behaviour that is usually much more intricate. More efficient RHG models can of course be obtained, provided the information is available, by a grouping on other factors beyond age and sex.

Suppose we replace the unknown θ_k in (6.4) by the estimates $\hat{\theta}_k$ obtained from the RHG model. This important special case of (6.6) is discussed in detail in Särndal *et al.* (1992, Chapter 15), where an appropriate variance estimator is also given. It has two components, one measuring the sampling variance, the other the nonresponse variance. The point estimator is (very nearly) unbiased if the assumed RHG model is a true representation of the response mechanism. But as already noted, no matter what groupings we can establish for the sample elements, they are not likely to perfectly meet the requirement of equal response probability of all elements within a group. Other attempts include logistic regression modelling and exponential modelling, as in Ekholm and Laaksonen (1991) and Folsom (1991). We further examine such possibilities in Section 8.4.

To summarize, the two-phase approach to weighting requires a modelling of the response mechanism as an initial step for the estimators (6.3) and (6.6). One must (i) give the mathematical formulation of the model, and (ii) select the explanatory variables for this model from a larger set of available auxiliary variables. This requires analysis and decision making.

6.2. AUXILIARY VECTORS AND AUXILIARY INFORMATION

The efficient use of auxiliary information is the key to reliable estimation in the presence of nonresponse. This holds both for the two-phase approach in the preceding section and for the *calibration approach to weighting* developed in this chapter and explored in the following chapters. The calibration approach is simpler and more direct than the two-phase approach, without jeopardizing the capacity to adjust for nonresponse. Many estimators in the two-phase approach are also derivable from the calibration approach.

Auxiliary information is transmitted by an auxiliary vector. The *auxiliary vector* is a vector whose value is known for every responding element, $k \in r$, and in addition there exists information on this vector for a *larger set* than r. This information will contribute both to protection against nonresponse and to variance reduction. Our generic notation for an auxiliary vector will be $\mathbf{x}$, and its value for element k is denoted $\mathbf{x}_k$.

The information for the larger set provides *information input*, necessary for the computation of calibrated weights.

For the calibration approach we distinguish three different cases, called InfoU, InfoS and InfoUS, depending on the information at hand. They are as follows:

InfoU. Information is available at the level of the population U. Denote by $\mathbf{x}_k^*$ a vector of dimension $J^* \geq 1$ such that:

(i) the population vector total $\sum_U \mathbf{x}_k^*$ is known;
(ii) for every $k \in r$, the vector value $\mathbf{x}_k^*$ is known.

The auxiliary vector for this case is $\mathbf{x}_k = \mathbf{x}_k^*$. We often refer to $\mathbf{x}_k^*$ as the 'star vector'.

InfoS. Information is available at the level of the sample s, but there is none at the level of the population. Denote by $\mathbf{x}_k^\circ$ a vector of dimension $J^\circ \geq 1$ such that:

(i) for every $k \in s$, the vector value $\mathbf{x}_k^\circ$ is known, whereas $\sum_U \mathbf{x}_k^\circ$ is unknown;
(ii) for every $k \in r$, the vector value $\mathbf{x}_k^\circ$ is known.

The auxiliary vector for this case is $\mathbf{x}_k = \mathbf{x}_k^\circ$. We often refer to $\mathbf{x}_k^\circ$ as the 'moon vector'. It carries information at a lower level than the star vector.

InfoUS. Both types of information, InfoU and InfoS, are available and used in combination for weight computation. One option is to formulate the auxiliary vector as the stacked vector $\mathbf{x}_k = \begin{pmatrix} \mathbf{x}_k^* \\ \mathbf{x}_k^\circ \end{pmatrix}$ of dimension $J^* + J^\circ$.

Conditions (i) and (ii) in InfoU and InfoS are *minimally required* for the point estimation procedures that we develop. Let us examine how InfoU and InfoS arise in practice. Possible sources of information are the survey itself, a census of the population in question, an administrative register, or a matching of several such registers.

First consider InfoU. Conditions (i) and (ii) prevail under two important scenarios found in practice: (a) The auxiliary vector total is 'imported' from a reliable source external to the survey itself. (b) There exists a register that lists the population elements from 1 to N, and such that a vector value $\mathbf{x}_k^*$ is specified for every element $k = 1, 2, \ldots, N$. The list may be the result of merging several administrative registers using a key or unique identifier. Examples include identity for persons, organizations or business enterprises. The list may also serve as a sampling frame (or to develop a frame) for the target population in the survey.

In the case of an *imported total*, the information $\sum_U \mathbf{x}_k^*$ comes from an external source, whereas the individual values $\mathbf{x}_k^*$ are known or measured in the survey for $k \in r$. It is then important that the vector behind the imported total $\sum_U \mathbf{x}_k^*$ in InfoU condition (i) measures the same concept as $\mathbf{x}_k^*$ in (ii). In other words, $\sum_U \mathbf{x}_k^*$ must not be an out-of-date or otherwise erroneous measure of the true population total of the vector $\mathbf{x}_k^*$ measured in the current survey for every $k \in r$.

Imported auxiliary totals are used for estimation in important surveys in North America and elsewhere. For example, the Canadian Labour Force Survey imports population figures for groups based on age category by sex by region (within a Canadian province). The external source is the census projection counts that are produced regularly as updates of the preceding quinquennial census of the population of Canada. Although not perfect, these projections are highly accurate. Thus the known information consists of the census projection counts for the different groups. The auxiliary vector for this case is $\mathbf{x}_k^* = (\gamma_{1k}, \ldots, \gamma_{pk}, \ldots, \gamma_{Pk})'$, where $\gamma_{pk} = 1$ for every k in group p and $\gamma_{pk} = 0$ for every k outside group p. The information $\sum_U \mathbf{x}_k^* = (N_1, \ldots, N_p, \ldots, N_P)'$ comes from the external source.

The vector $\mathbf{x}_k^*$ may also contain variables other than the categorical ones, provided the corresponding population totals can be imported from a reliable source.

A nuisance factor intervenes when the values $\mathbf{x}_k^*$ delivered by respondents contain measurement errors. This is particularly troublesome if such errors are not random but systematic, as may be the case with a sensitive variable. As an example, suppose that the sampled individuals are asked to quantify their alcohol consumption during a given month. There may be a tendency to understate one's consumption. As for a population total, suppose we can import the monthly sales figure from a source considered reliable, such as a state monopoly on alcoholic beverages. Even though this may give a reasonably accurate figure for total consumption, the trouble is that the measurements obtained from respondents are systematically on the low side. This disparity might severely perturb the weights computed by calibration.

Consider now the case of an existing *population list* used to satisfy the InfoU requirements. This situation is typical in surveys of individuals and households in several European countries, notably in Scandinavia, where a total population register lists all individuals k in the population (the frame) U. The coverage errors of these registers are minor.

The register may be the result of matching several administrative sources using a unique element identifier, such as the person identity number in the Scandinavian countries.

A total population register typically contain values of a variety of variables. The variable values are thus available for all elements in the population, not only for sampled elements. Examples of such variables are age and sex, employment status, region and type of residence, type and duration of formal education, and salary. As a result, a star vector value $\mathbf{x}_k^*$ can be specified for every individual $k \in U$. Thus $\mathbf{x}_k^*$ is known for every $k \in s$, as it is for every $k \in r$. By summing the values $\mathbf{x}_k^*$ on the frame we arrive at the desired total $\sum_U \mathbf{x}_k^*$, thereby fulfilling requirement (i).

We note that matching of registers can be avoided by *calibration on margins*, as explained in Section 7.6. In this procedure the auxiliary vector will not reflect interactions between explanatory factors, only the main effects. This may be a small price to pay for the simplification.

Now consider InfoS, characterized by a moon vector value $\mathbf{x}_k^\circ$ measured or otherwise known for every sampled element, $k \in s$. It is especially important to include in $\mathbf{x}_k^\circ$ variables that explain the nonresponse behaviour. Ingenuity and good judgement help the identification of such variables. An example is the

Table 6.1 Formulation of the auxiliary vector and the corresponding information input for the cases InfoU, InfoS and InfoUS.

Case	Auxiliary vector, $\mathbf{x}_k$	Information input, $\mathbf{X}$
InfoU	$\mathbf{x}_k^*$	$\mathbf{X}^* = \sum_U \mathbf{x}_k^*$
InfoS	$\mathbf{x}_k^\circ$	$\hat{\mathbf{X}}^\circ = \sum_s d_k \mathbf{x}_k^\circ$
InfoUS	$\begin{pmatrix} \mathbf{x}_k^* \\ \mathbf{x}_k^\circ \end{pmatrix}$	$\begin{pmatrix} \mathbf{X}^* \\ \hat{\mathbf{X}}^\circ \end{pmatrix}$

identity of the interviewer to whom element $k \in s$ is assigned. Other examples are given in Section 7.3.

One source of auxiliary variable values for $k \in s$ is the basic question method proposed by Bethlehem and Kersten (1985). It is motivated by the experience that persons who refuse to respond to an entire questionnaire may nevertheless willingly respond to one or two 'basic questions'. The basic question variables should reflect, or be related to, themes that are central to the survey in question, so that they correlate well with important survey variables, and they should also correlate well with response propensity.

Bethlehem and Kersten (1985) maintain that the basic question method is not limited to refusals. Responses to the basic questions may also be obtained by mail or telephone inquiry. Basic question variables can be entered into the vector $\mathbf{x}_k^\circ$, assuming that responses to those variables have been obtained to all, or virtually all, elements $k \in s$.

The notation for InfoU, InfoS and InfoUS is summarized in Table 6.1, where $\mathbf{X}$ is the generic notation for the information input that accompanies the auxiliary vector $\mathbf{x}_k$.

We can view InfoU as the special case of InfoUS obtained when there exists a star vector but no moon vector; the auxiliary vector thus reduces to $\mathbf{x}_k = \mathbf{x}_k^*$. Similarly, InfoS is the special case of InfoUS obtained when there is no star vector, so that $\mathbf{x}_k = \mathbf{x}_k^\circ$.

The information input $\hat{\mathbf{X}}^\circ = \sum_s d_k \mathbf{x}_k^\circ$ is the unbiased Horvitz–Thompson estimate of the unknown total $\mathbf{X}^\circ = \sum_U \mathbf{x}_k^\circ$. If available, a better estimate (but unbiased or nearly so) is allowed to take the place of $\sum_s d_k \mathbf{x}_k^\circ$. One such alternative is of the GREG type, $\hat{\mathbf{X}}^\circ = \sum_s d_k g_k \mathbf{x}_k^\circ$, where g_k is given by (4.10). This requires more information on the part of the star vector than what InfoU specifies at the beginning of this section, namely, that $\mathbf{x}_k^*$ must be known individually for every $k \in s$.

6.3.   THE CALIBRATION APPROACH: SOME TERMINOLOGY

Our main concern is with surveys affected by nonresponse. In such cases, desirable properties of an estimator of $Y = \sum_U y_k$ are:

(i)   a small bias;
(ii)   a small variance;

(iii)   a weight system that reproduces the auxiliary information input when applied to the auxiliary variables;

(iv)   a weighting system that is useful for estimating the total of every y-variable in a multi-purpose survey.

Property (i) is particularly important, and the prospects for achieving a small bias are analysed at considerable length in Chapters 9 and 10. As for the variance, it will consist of the sum of a sampling variance component and a non-response variance component. Ideally both components should be small. The property expressed by (iii) is the cornerstone in the calibration approach. Property (iv) expresses the desirability of a unique weighting system. This aspect is important for routine and timely production of statistics, especially in large government surveys.

We now describe the *calibration approach* to weighting for surveys with non-response. The first step is to fix a suitable auxiliary vector by a selection of variables from a larger set of available variables. To guide this selection we formulate a few simple principles in Section 10.2. The next step is the computation of the *calibrated weights*, using existing computer software. Several existing software can be used for this task, and some also carry out variance estimation for particular sampling designs.

The calibration approach produces a *calibration estimator* of Y, denoted $\hat{Y}_W$, and a corresponding *variance estimator*, denoted $\hat{V}(\hat{Y}_W)$. The index W is used to suggest 'weighting'. The approach is very general. In the presence of powerful auxiliary information, it meets the double objective of reducing the sampling error and the nonresponse error. It can be applied for any of the common sampling designs and for any auxiliary vector. The approach does not appeal explicitly to any model(s). It is well suited to routine and timeless production of estimates, for example, in a national statistical agency. In the following we highlight both the theoretical and the practical aspects of the approach. The theoretical development is influenced by, among others, Lundström and Särndal (1999) and Deville (2000).

6.4.   POINT ESTIMATION UNDER THE CALIBRATION APPROACH

We now present the *calibration estimator* of $Y = \sum_U y_k$, denoted $\hat{Y}_W$. The calibrated weights are denoted w_k for $k \in r$. The calibration estimator is given by

$$\hat{Y}_W = \sum_r w_k y_k. \tag{6.7}$$

It depends on the strength of the auxiliary information whether or not the w_k can successfully protect against nonresponse bias. We focus on the most general case, InfoUS, where the auxiliary vector is $\mathbf{x}_k = \begin{pmatrix} \mathbf{x}_k^* \\ \mathbf{x}_k^\circ \end{pmatrix}$ of dimension $J^* + J^\circ$ and the information input is $\mathbf{X} = \begin{pmatrix} \mathbf{X}^* \\ \hat{\mathbf{X}}^\circ \end{pmatrix}$. Special cases are InfoU, for which

$\mathbf{x}_k = \mathbf{x}_k^*$, and InfoS, for which $\mathbf{x}_k = \mathbf{x}_k^\circ$. The three cases lead to different calibrated weight systems.

We seek a weight system w_k for $k \in r$ that satisfies the *calibration equation*

$$\sum_r w_k \mathbf{x}_k = \mathbf{X}. \qquad (6.8)$$

Equation (6.8) implies for $\mathbf{x}_k^*$ that $\sum_r w_k \mathbf{x}_k^* = \sum_U \mathbf{x}_k^*$, and for $\mathbf{x}_k^\circ$ that $\sum_r w_k \mathbf{x}_k^\circ = \sum_s d_k \mathbf{x}_k^\circ$. Weights that satisfy (6.8) are said to be *calibrated* to the information input $\mathbf{X}$. They will exactly reproduce the given information $\mathbf{X}$ when applied to the auxiliary vector values $\mathbf{x}_k$ and summed over the response set r. With another often used term, the weight system w_k for $k \in r$ is said to be *consistent* with the information $\mathbf{X}$.

As a result of the sample selection, element k carries weight $d_k = 1/\pi_k$. When there is nonresponse, the weights d_k for $k \in r$ are by themselves too small on average to produce acceptable estimates. The weighted sum $\sum_r d_k y_k$ is a clear underestimate of $Y = \sum_U y_k$. The d_k must be raised. We seek new weights that are greater than the d_k for all, or at least for a majority, of the responding elements. Denote by v_k the factor, or weight, by which we multiply d_k to obtain the new weight for element k. That is, $w_k = d_k v_k$.

What form should the weight v_k take? It should reflect the known individual characteristics of the element $k \in r$, summarized by the vector value $\mathbf{x}_k$. One simple form is then $v_k = 1 + \boldsymbol{\lambda}' \mathbf{x}_k$, where $\boldsymbol{\lambda}$ is a vector to be determined. In this formulation, we let v_k depend linearly on the known value $\mathbf{x}_k$ for element k. We comment later on other formulations.

It is now an easy exercise to determine $\boldsymbol{\lambda}$ to satisfy the calibration requirement. Insert $v_k = 1 + \boldsymbol{\lambda}' \mathbf{x}_k$ in (6.8) and solve for $\boldsymbol{\lambda}'$. We get $\boldsymbol{\lambda}' = \boldsymbol{\lambda}_r'$, where

$$\boldsymbol{\lambda}_r' = \left(\mathbf{X} - \sum_r d_k \mathbf{x}_k \right)' \left(\sum_r d_k \mathbf{x}_k \mathbf{x}_k' \right)^{-1} \qquad (6.9)$$

assuming that the inverse of the matrix $\sum_r d_k \mathbf{x}_k \mathbf{x}_k'$ exists. The new weight is $w_k = d_k + d_k \boldsymbol{\lambda}_r' \mathbf{x}_k$, where the added term $d_k \boldsymbol{\lambda}_r' \mathbf{x}_k$ will be positive for most (but not necessarily all) elements. This term expands the insufficiently large sampling weight d_k to a more reasonable, usually higher value.

We have now determined weights that account for nonresponse and are calibrated to the given information. They are given by

$$w_k = d_k v_k, \qquad v_k = 1 + \boldsymbol{\lambda}_r' \mathbf{x}_k \qquad (6.10)$$

where $\boldsymbol{\lambda}_r' = \left(\mathbf{X} - \sum_r d_k \mathbf{x}_k \right)' \left(\sum_r d_k \mathbf{x}_k \mathbf{x}_k' \right)^{-1}$. The resulting calibration estimator is

$$\hat{Y}_W = \sum_r w_k y_k. \qquad (6.11)$$

Apart from the information input $\mathbf{X}$, the computation of the w_k requires two sums over the response set r. Both sums can be computed because $\mathbf{x}_k$ is known for every $k \in r$. The matrix $\left(\sum_r d_k \mathbf{x}_k \mathbf{x}_k' \right)$ requiring inversion is of dimension $(J^* + J^\circ) \times (J^* + J^\circ)$, where J^* and J° are the respective dimensions of $\mathbf{x}_k^*$ and $\mathbf{x}_k^\circ$.

We shall call the weighting defined by (6.10) the *standard weighting*. Alternative calibrated weightings are discussed in Section 6.7.

Example 6.1.   The simplest auxiliary vector

The simplest auxiliary vector is one that is constant for all k. Since it recognizes no individual differences, it is inefficient for nonresponse treatment. But it gives the most basic example of calibration estimation. With $\mathbf{x}_k = \mathbf{x}_k^* = 1$ for all k, we obtain from (6.10) $w_k = d_k \left(N / \sum_r d_k \right)$ and $\hat{Y}_W = N\bar{y}_{r;d}$, where $\bar{y}_{r;d} = \left(\sum_r d_k y_k / \sum_r d_k \right)$ is the design-weighted y-mean for the respondents. In particular, for SI, $w_k = N/m$ for all k and $\hat{Y}_W = N\bar{y}_r$, where $\bar{y}_r = \sum_r y_k / m$ and m is the size of the response set r. Examples of calibration estimators based on more useful $\mathbf{x}_k$-vectors are given in Chapter 7.   □

Remark 6.1

For a given vector $\mathbf{x}_k$ and corresponding information input $\mathbf{X}$, the computation of w_k defined by (6.10) can yield a few negative or unduly large values v_k. Negative weights are considered by many as undesirable, and so are unduly large positive weights. There are different ways to address this issue. Some of the existing software has built-in procedures that allow the weights to be restricted to within prespecified intervals. Preferably one should keep the issue at the front of one's mind when constructing the auxiliary vector. Prior to a final decision on $\mathbf{x}_k$, one should make sure that the vector does not contain x-variables likely to cause the problem with undesirable weights. We return to this issue several times in the following, especially in Chapters 10 and 11.   □

Remark 6.2

We have argued that the linear form, $v_k = 1 + \boldsymbol{\lambda}'\mathbf{x}_k$, is reasonable for the adjustment. Nonlinear forms can also be considered, as in Deville (2000). We could postulate $v_k = F(\boldsymbol{\lambda}'\mathbf{x}_k)$ for some function $F(\cdot)$ with suitable properties, for example, $F(\boldsymbol{\lambda}'\mathbf{x}_k) = \exp(\boldsymbol{\lambda}'\mathbf{x}_k)$. We would then seek a vector $\boldsymbol{\lambda}$ to satisfy the calibration equation $\sum_r d_k F(\boldsymbol{\lambda}'\mathbf{x}_k) = \mathbf{X}$. If that vector is $\boldsymbol{\lambda}'_r$, the calibrated weights are $w_k = d_k F(\boldsymbol{\lambda}'_r \mathbf{x}_k)$. For our purposes, the linear form will suffice; it has considerable computational advantages.   □

6.5.   CALIBRATION ESTIMATORS FOR DOMAINS

Many surveys, including large government surveys, require estimation not only for the whole population but also for various domains or subpopulations. Let U_q be any domain of U, $U_q \subseteq U$. We seek to estimate the domain y-total, $Y_q = \sum_{U_q} y_k = \sum_U y_{qk}$, where y_{qk} is the value for element k of the domain-specific study variable y_q, defined by $y_{qk} = y_k$ for $k \in U_q$ and $y_{qk} = 0$ for $k \in U - U_q$. We now propose a procedure for estimating Y_q when the auxiliary vector is $\mathbf{x}_k$ and the information input is $\mathbf{X}$, as in Table 6.1. The calibrated weights for that information are given by (6.10). Consequently, the calibration estimator of the domain total Y_q is given by

$$\hat{Y}_{q\mathrm{W}} = \sum\nolimits_r w_k y_{qk}. \tag{6.12}$$

In many surveys, interest is focused on a set of domains $U_1, \ldots, U_q, \ldots, U_Q$ that form a partition of U, as is often the case when the Q domains are regions that make up a country. We apply the same set of calibrated weights for all domains, and obtain the estimates $\hat{Y}_{q\mathrm{W}} = \sum_r w_k y_{qk}$ for $q = 1, \ldots, Q$. There is one domain-specific variable y_q for every domain. The weights remain the same for all domains. An attractive property of this procedure is that the domain estimates $\hat{Y}_{1\mathrm{W}}, \ldots, \hat{Y}_{q\mathrm{W}}, \ldots, \hat{Y}_{Q\mathrm{W}}$ will automatically add up to the calibration estimate for the whole population, namely $\hat{Y}_\mathrm{W} = \sum_r w_k y_k$. The property follows from

$$\sum_{q=1}^{Q} \hat{Y}_{q\mathrm{W}} = \sum_{q=1}^{Q} \sum\nolimits_r w_k y_{qk} = \sum\nolimits_r w_k \sum_{q=1}^{Q} y_{qk} = \sum\nolimits_r w_k y_k = \hat{Y}_\mathrm{W}.$$

All domain estimates are based on the information input that is available in the survey. Unless otherwise stated, we shall assume that estimation for one or more domains is handled by the principle expressed by (6.12). Estimation for a domain, provided it contains sufficient y-data, is then a simple matter. We need not discuss it extensively in the chapters that follow.

As property (iii) in Section 3.3 suggests, the auxiliary vector should identify the most important domains. Suppose x is an auxiliary variable well correlated with y, and that we seek to estimate the y-total for the domain U_q, $Y_q = \sum_{U_q} y_k$. The known x-total for the whole population, $\sum_U x_k$, is a piece of information that may be of limited value only for estimating the y-total of a domain situated substantially below the level of the whole population. Its effectiveness for reducing bias and variance may be very modest for a not-so-large region of the whole country (unless x can effectively explain the response propensity throughout the population, in accordance with property (i) in Section 3.3). By contrast, consider the auxiliary variable that identifies the domain, $x_{qk} = \delta_{qk} x_k$, where δ_{qk} is the domain identifier (4.3). The associated information input $\sum_U x_{qk} = \sum_{U_q} x_k$, if available, stands a much better chance of reducing bias and variance than the information $\sum_U x_k$.

6.6. COMMENTS ON THE CALIBRATION APPROACH

Early in Section 6.3, four desirable estimator properties were formulated. Are these properties present in the calibration estimator $\hat{Y}_\mathrm{W} = \sum_r w_k y_k$ with the weights given by (6.10)? Properties (i) and (ii) relating to bias and variance are thoroughly discussed in the following chapters. In particular, Chapters 9 and 10 focus on the bias and methods to control it. These methods require a powerful auxiliary vector $\mathbf{x}_k$, and we shall examine techniques for constructing such a vector. Property (iii) is satisfied by construction. Property (iv) expresses the desirability of a set of weights that are applicable, with good results, to *all* variables of interest in a survey. The calibrated weights w_k meet this objective in that they incorporate the information considered the best possible under the conditions of the survey.

The following remark points out another interesting property of the calibration estimator.

Remark 6.3

Consider the case InfoU, where $\mathbf{x}_k = \mathbf{x}_k^*$ and $\mathbf{X} = \mathbf{X}^* = \sum_U \mathbf{x}_k^*$. Assume that a perfect linear relationship exists in the population between the study variable y_k and the auxiliary vector $\mathbf{x}_k^*$, so that, for every $k \in U$,

$$y_k = (\mathbf{x}_k^*)'\boldsymbol{\beta}^* \tag{6.13}$$

where $\boldsymbol{\beta}^*$ is a column vector of (unknown) constants. Then the calibration estimator $\hat{Y}_W$ gives an exact estimate of the target total Y. This follows from

$$\hat{Y}_W = \sum_r w_k y_k = \left(\sum_r w_k \mathbf{x}_k^*\right)' \boldsymbol{\beta}^* = \left(\sum_U \mathbf{x}_k^*\right)' \boldsymbol{\beta}^* = \sum_U y_k = Y.$$

All that is needed to establish the equality $\hat{Y}_W = Y$ is the calibration property of the weights, given by (6.8). ☐

In practice, the perfect linear relationship expressed by (6.13) does not hold. If it did, there would hardly be a need to make that y-variable part of the survey. But the result does suggest that when a strong linear relationship exists between y_k and $\mathbf{x}_k^*$, the calibration estimator $\hat{Y}_W$ should come close in value to the target Y. Both the sampling error and the nonresponse error should be essentially eliminated.

We mention some additional features of the calibration approach:

- *Generality.* Although $\hat{Y}_W$ is called 'the calibration estimator', it is in reality the expression for a wide family of estimators, corresponding to the different formulations of $\mathbf{x}_k$. Great generality and flexibility are obtained from the fact that the $\mathbf{x}_k$-vector in (6.10) can have virtually any form, as long as the corresponding information input $\mathbf{X}$ is available, and as long as the matrix is invertible.
- *Computational aspects.* When the calibration approach is used in practice, there is no need for algebraic derivation of an estimator formula for a chosen $\mathbf{x}_k$-vector. Instead the focus is on the computation, carried out by existing software. We can rely on a number of available software programs. The user needs to specify the $\mathbf{x}_k$-vector and the corresponding information input. The software will produce the calibrated weights w_k and the corresponding calibration estimate $\hat{Y}_W = \sum_r w_k y_k$. Some software also computes the associated variance.
- *Conventional techniques.* There exist techniques for nonresponse weighting that can be termed 'conventional', considering the attention that they have attracted over the years in the literature. It is convincing to find that many of these are obtainable as simple special cases of the calibration estimator

$\hat{Y}_W$. In Chapter 7 we derive some of these conventional techniques as special cases of $\hat{Y}_W$. They correspond to basic formulations of the $\mathbf{x}_k$-vector.

6.7. ALTERNATIVE SETS OF CALIBRATED WEIGHTS

Calibrated weights are not unique. The calibration equation (6.8) poses only a weak constraint on the weights. As we now note, there exist many sets of calibrated weights, for a given $\mathbf{x}_k$-vector with given information input $\mathbf{X}$.

We need some additional concepts in order to see the full width of the calibration technique: *initial weights, final weights* and *instrument vector*. The d_k are initial weights in (6.10), in the sense that the computation defined by (6.10) will transform the d_k into the new weights w_k, which are the final weights.

First note that we can work with other initial weights than the d_k and still satisfy the requirement that the final weights calibrate to the given information $\mathbf{X}$. Let $d_{\alpha k}$, for $k \in r$, be any set of positive weights. In (6.10) and in the corresponding vector λ'_r, replace d_k by $d_{\alpha k}$. The resulting final weights w_k will still satisfy the calibration equation (6.8). For example, we could consider $d_{\alpha k} = C d_k$, where C is a positive value not depending on k, such as n/m, the inverse of the overall response rate in the survey.

We now note a second replacement in (6.10) that will also leave the calibration property (6.8) unchanged. Let $\mathbf{z}_k$ be any vector value specified for $k \in r$ and with the same dimension as $\mathbf{x}_k$. The vector $\mathbf{z}_k$ can be a specified function of $\mathbf{x}_k$ or of other background data about k. In (6.10), replace $v_k = 1 + \lambda'_r \mathbf{x}_k$ and $\left(\sum_r d_k \mathbf{x}_k \mathbf{x}'_k \right)^{-1}$ by $v_k = 1 + \lambda'_r \mathbf{z}_k$ and $\left(\sum_r d_k \mathbf{z}_k \mathbf{x}'_k \right)^{-1}$, respectively. The resulting new weights are still calibrated to the information input $\mathbf{X}$. We call $\mathbf{z}_k$ an *instrument vector* for the calibration.

Using the concepts of initial weights and instrument vector, we get the calibrated weight system

$$w_k = d_{\alpha k} v_k, \quad v_k = 1 + \lambda'_r \mathbf{z}_k \tag{6.14}$$

where $\lambda'_r = \left(\mathbf{X} - \sum_r d_{\alpha k} \mathbf{x}_k \right)' \left(\sum_r d_{\alpha k} \mathbf{z}_k \mathbf{x}'_k \right)^{-1}$. These w_k satisfy the calibration equation (6.8) for *any* positive initial weights $d_{\alpha k}$ and *any* instrument $\mathbf{z}_k$, as long as the matrix $\sum_r d_{\alpha k} \mathbf{z}_k \mathbf{x}'_k$ can be inverted. Using these weights, the calibration estimator is $\hat{Y}_W = \sum_r w_k y_k$.

The *standard weighting* defined by (6.10) is the special case of (6.14) obtained for $d_{\alpha k} = d_k$ and $\mathbf{z}_k = \mathbf{x}_k$. Unless otherwise stated, we assume in the following that the *standard specifications* $d_{\alpha k} = d_k$ and $\mathbf{z}_k = \mathbf{x}_k$ apply. They have an appealing simplicity, and the following chapters show that they work well.

In summary, to compute the weights w_k given by (6.14) we need to specify three types of entities:

- the initial weights $d_{\alpha k}$;
- the auxiliary vector values $\mathbf{x}_k$ and the corresponding information input $\mathbf{X}$;
- the instrument vector values $\mathbf{z}_k$, if different from $\mathbf{x}_k$.

The calibration procedure transforms the initial weights via an input of information into final weights given by (6.10) or more generally by (6.14). The final weights are also called *calibrated weights*, and they take the nonresponse into account in an appropriate fashion.

Remark 6.4

A desirable (but not mandatory) property of the calibrated weights is that they add up to the population size N. When they do, the weight system will correctly estimate the population size. It is a justified requirement, but by itself it gives little guidance for the choice of $d_{\alpha k}$ and $\mathbf{z}_k$. A great many calibrated weights systems have this property. Regardless of the initial weights $d_{\alpha k}$ and the instrument vector $\mathbf{z}_k$ in (6.14), the equation $\sum_r w_k = N$ holds for InfoU and InfoUS whenever the star vector $\mathbf{x}_k^*$ contains the constant 1 for all k. It is not a serious drawback if $\sum_U w_k = N$ does not hold, considering, for example, that in some surveys, the population size N is unknown. ☐

Because of the freedom to choose $d_{\alpha k}$ and $\mathbf{z}_k$, there exist many weight systems calibrated to a given input of information, $\mathbf{X}$. To every such weight system there corresponds a calibration estimator $\hat{Y}_W = \sum_r w_k y_k$. This raises the question of how to choose $d_{\alpha k}$ and $\mathbf{z}_k$. We shall most often use $d_{\alpha k} = d_k$ as initial weights. One reason is that the final weights w_k are in many cases invariant to a preliminary adjustment of the d_k, such as $d_{\alpha k} = C d_k$, where C is a value not dependent on k, for example, $C = n/m$, the inverse of the overall response rate. We show this invariance in the next section.

Now consider the choice of $\mathbf{z}_k$. In the majority of all cases we take $\mathbf{z}_k$ to be identical to $\mathbf{x}_k$, *the standard choice*. It simplifies the reading of the following sections to always keep this in mind. We consider a few cases where $\mathbf{z}_k$ is different from $\mathbf{x}_k$, allowing us in Chapter 7 to identify some well-known estimators as special cases of the general expression $\hat{Y}_W = \sum_r w_k y_k$. But in practice, we can specify the components of $\mathbf{z}_k$ as other functions of the components of $\mathbf{x}_k$. If $\mathbf{x}_k = (x_{1k}, x_{2k})'$, where x_{1k} and x_{2k} are positive, we can obtain calibrated weights from (6.14) by taking, for example, $\mathbf{z}_k = (\sqrt{x_{1k}}, \sqrt{x_{2k}})'$. We do not imply that a $\mathbf{z}_k$ different from $\mathbf{x}_k$ gives weights in some sense better than those for $\mathbf{z}_k = \mathbf{x}_k$, nor do we recommend any specific form of $\mathbf{z}_k$. We simply wish to alert the reader to a generality of the calibration approach that can perhaps be explored and used to advantage in certain settings.

Remark 6.5

For complete response, a close relation exists between the calibration estimator $\hat{Y}_W = \sum_r w_k y_k$ with weights defined by (6.14) and the GREG estimator (4.9). If $r = s$ and if $\mathbf{x}_k = \mathbf{x}_k^*$, $\mathbf{z}_k = c_k \mathbf{x}_k^*$, the two estimators are identical. This is an attractive property, because the GREG estimator is known to have favourable properties: a bias very near zero and a small variance when y_k is well explained by $\mathbf{x}_k^*$. When the nonresponse is very limited, the calibration estimator is thus nearly a GREG estimator. ☐

Remark 6.6

The two-phase estimator $\hat{Y} = \sum_r (d_k/\hat{\theta}_k) g_{k\hat{\theta}} y_k$ given by (6.6) is a calibration estimator with weights of the form (6.14). The initial weights are $d_{\alpha k} = d_k/\hat{\theta}_k$, the auxiliary vector is $\mathbf{x}_k = \mathbf{x}_k^*$, the instrument is $\mathbf{z}_k = \mathbf{x}_k^*$, the information input is $\sum_U \mathbf{x}_k^*$, and the final weights are

$$w_k = \frac{d_k}{\hat{\theta}_k} \left\{ 1 + \left(\sum_U \mathbf{x}_k^* - \sum_r \frac{d_k}{\hat{\theta}_k} \mathbf{x}_k^* \right)' \left(\sum_r \frac{d_k}{\hat{\theta}_k} \mathbf{x}_k^* (\mathbf{x}_k^*)' \right)^{-1} \mathbf{x}_k^* \right\}.$$

It is easy to see that they satisfy the calibration equation (6.8) with $\mathbf{X} = \sum_U \mathbf{x}_k^*$.
□

Remark 6.7

Even if $\sum_r d_{\alpha k} \mathbf{z}_k \mathbf{x}_k'$ is singular, calibrated weights could still be given by (6.14) if we substitute a *generalized inverse*; however, we need not consider this possibility here.
□

6.8. PROPERTIES OF THE CALIBRATED WEIGHTS

To get a further perspective on the calibration process, let us take a closer look at the weights w_k given by (6.14). We can write w_k as

$$w_k = w_{Mk} + w_{Rk}$$

where w_{Mk} is the 'main component' of w_k given by

$$w_{Mk} = d_{\alpha k} \left\{ \mathbf{X}' \left(\sum_r d_{\alpha k} \mathbf{z}_k \mathbf{x}_k' \right)^{-1} \mathbf{z}_k \right\}$$

and w_{Rk} is the 'remainder component' given by

$$w_{Rk} = d_{\alpha k} \left\{ 1 - \left(\sum_r d_{\alpha k} \mathbf{x}_k \right)' \left(\sum_r d_{\alpha k} \mathbf{z}_k \mathbf{x}_k' \right)^{-1} \mathbf{z}_k \right\}.$$

We note that $\sum_r w_{Mk} \mathbf{x}_k = \mathbf{X}$ and $\sum_r w_{Rk} \mathbf{x}_k = \mathbf{0}$. That is, the desired calibration to the information input $\mathbf{X}$ is achieved already by the w_{Mk}. Adding w_{Rk} to w_{Mk} does not compromise the calibration property.

Suppose the sampling weights are initial weights, so that $d_{\alpha k} = d_k$. In the presence of nonresponse, the d_k are too small on average for estimating the population total $Y = \sum_U y_k$. That is, $\sum_r d_k y_k$ could severely underestimate Y. But the factor $\mathbf{X}' \left(\sum_r d_k \mathbf{z}_k \mathbf{x}_k' \right)^{-1} \mathbf{z}_k$ in w_{Mk} expands d_k to a more appropriate level.

The remainder weight component w_{Rk} contributes little or nothing to w_k. If not equal to zero, w_{Rk} is usually numerically small compared to w_{Mk}. In particular, $w_{Rk} = 0$ for all k, and thus $w_k = w_{Mk}$, if we choose the instrumental vector $\mathbf{z}_k$ so that, for some constant vector $\boldsymbol{\mu}$ not dependent on k, $\boldsymbol{\mu}' \mathbf{z}_k = 1$ holds for all k, because then $\sum_r d_{\alpha k} \mathbf{x}_k = \boldsymbol{\mu}' \sum_r d_{\alpha k} \mathbf{z}_k \mathbf{x}_k'$ and

$$w_{Rk} = 1 - \left(\boldsymbol{\mu}' \sum_r d_{\alpha k} \mathbf{z}_k \mathbf{x}_k' \right) \left(\sum_r d_{\alpha k} \mathbf{z}_k \mathbf{x}_k' \right)^{-1} \mathbf{z}_k = 1 - \boldsymbol{\mu}' \mathbf{z}_k = 0$$

for all k. In a majority of the applications to be discussed, it is possible to find a vector $\boldsymbol{\mu}$ to satisfy the condition $\boldsymbol{\mu}'\mathbf{z}_k = 1$ for all k. Then the calibrated weights are simply $w_k = w_{Mk}$.

Remark 6.8

The condition on $\mathbf{z}_k$ that we have just noted will recur in many places in the following. The condition is: for the vector $\mathbf{z}_k$, there exists a constant (that is, non-random) vector $\boldsymbol{\mu}$ not dependent on k such that $\boldsymbol{\mu}'\mathbf{z}_k = 1$ for all $k \in U$. It is not a highly restrictive condition. It is satisfied in many of the cases we consider. For brevity, we shall from now on refer to the condition as '$\boldsymbol{\mu}'\mathbf{z}_k = 1$ holds for all k'. Because $\mathbf{z}_k = \mathbf{x}_k$ in a majority of cases that we discuss, the condition will often read '$\boldsymbol{\mu}'\mathbf{x}_k = 1$ holds for all k'. □

To illustrate the weight components w_{Mk} and w_{Rk}, consider SI with sample size n. The sampling weights are $d_k = N/n$ for all k. They need to be expanded if $m < n$, where m is the number of respondents. Consider the simplest of conditions, so that $\mathbf{x}_k = \mathbf{x}_k^* = \mathbf{z}_k = 1$ and $\mathbf{X} = \sum_U \mathbf{x}_k^* = N$. The expansion factor in w_{Mk} is then $\mathbf{X}' \left(\sum_r d_k \mathbf{x}_k \mathbf{x}_k' \right)^{-1} \mathbf{z}_k = n/m$ for all k, so $w_{Mk} = (N/n)(n/m) = N/m$. Furthermore, $w_{Rk} = 0$ for all k. Hence the starting weights $d_k = N/n$ are expanded by calibration to a reasonable higher level, $w_k = N/m$, and we get $\hat{Y}_W = N\bar{y}_r$, where $\bar{y}_r$ is the respondent mean of y_k. This estimator may not protect adequately for nonresponse bias, because $\mathbf{x}_k = \mathbf{x}_k^* = 1$ is a very weak auxiliary vector, but at least the estimate will be placed at a 'correct level'.

Another property of the main component w_{Mk} is easy to see: multiplying the initial weights by a constant factor leaves the w_{Mk} unchanged. The same w_{Mk} are obtained with the initial weights $d_{\alpha k} = d_k$ as with the initial weights $d_{\alpha k} = C d_k$, where C is any positive constant not dependent on any individual k, such as $C = n/m$ or $C = \left(\sum_s d_k \right) / \left(\sum_r d_k \right)$. This invariance is important, because some would argue that more appropriate initial weights than $d_{\alpha k} = d_k$ would be $d_{\alpha k} = (n/m)d_k$ or $d_{\alpha k} = \left(\sum_s d_k \right) / \left(\sum_r d_k \right) d_k$. In both cases we have multiplied the insufficiently large design weights d_k by a positive constant to set the initial weights at a more correct level. But the multiplicative factor C has no impact on the final w_{Mk}. Justified by this invariance under multiplication by a common value, we shall frequently use $d_{\alpha k} = d_k$ as initial weights.

To illustrate, let us again consider SI. Some would argue that instead of $d_k = N/n$, the calibration should use $d_{\alpha k} = (n/m)(N/n) = N/m$ as initial weights because they have the 'correct level'. But the counter argument is that w_{Mk} remains unchanged, so we might as well start with $d_k = N/n$, even though they are 'too small'.

A sampling design of great importance in practice is STSI. With notation as in Example 4.1, the sampling weights are $d_k = N_h/n_h$ for all k in stratum h. In the direct approach, we would use the vector $\mathbf{z}_k = \mathbf{x}_k$ of our choice, and $d_{\alpha k} = d_k = N_h/n_h$ as initial weights, to obtain final calibrated weights from (6.14). An alternative approach is to pose 'initial weights at a correct level'. A commonly used method is to expand the sampling weights stratum by stratum, by the inverse response rate in each stratum, and then to obtain final calibrated weights by (6.14). It depends on the $\mathbf{x}_k$-vector whether the two approaches give identical final weights. We further examine the question in Remark 7.2.

We summarize some of our findings in this section in the following remark:

Remark 6.9

Suppose that $\boldsymbol{\mu}'\mathbf{z}_k = 1$ holds for all k. Then the calibrated weights (6.14) simplify to

$$w_k = w_{\mathrm{M}k} = \mathbf{X}' \left(\sum_r d_{\alpha k}\mathbf{z}_k\mathbf{x}'_k \right)^{-1} d_{\alpha k}\mathbf{z}_k. \tag{6.15}$$

Furthermore, the same calibrated weights w_k are obtained for all initial weight systems of the form $d_{\alpha k} = Cd_k$, where $d_k = 1/\pi_k$ and C is any positive value not dependent on k. $\qquad\square$

CHAPTER 7

Examples of Calibration Estimators

7.1. EXAMPLES OF FAMILIAR ESTIMATORS FOR DATA WITH NONRESPONSE

Many survey statisticians are accustomed to specific nonresponse-adjusted estimators. These 'familiar formulae' have special names. Each named estimator is viewed as a particular 'method' for nonresponse treatment, as if it had its own unique roots. In reality the only difference lies in the extent of the auxiliary information.

The calibration approach in Chapter 6 invites us to take a broader perspective. There is just one general approach. It generates a wide family of estimators $\hat{Y}_W$. Its members correspond to different inputs of information. In practice, the formulae for special cases of $\hat{Y}_W$ need never require the attention of the practising statistician. Instead, he or she will focus on how to compose a powerful auxiliary vector $\mathbf{x}_k$. After that, all the necessary computation is carried out by existing software.

The objective of this section is to derive and confirm some of the formulae that many readers are familiar with. We obtain them as special cases of the calibration approach. We examine only very few of all the possible calibration estimators.

The chapter is arranged as follows. We consider simple specifications of the auxiliary vector and the associated information input. Under these specifications, we derive the explicit form of $\hat{Y}_W$. Unless otherwise stated, the standard weighting (6.10) is used. A few instances require the more general weighting (6.14).

We assume that sampling from the population U of size N is carried out with a sampling design with the sampling weights $d_k = 1/\pi_k$. Having found $\hat{Y}_W$ for a general design, it is straightforward to find its expression for particular designs. We consider in particular SI of n elements chosen from N, because the formulae are usually easy to understand in this case, where $d_k = N/n$ for all $k \in U$. We also consider STSI because of its great importance in practice; in that case, the design weights are $d_k = N_h/n_h$ for all k in stratum h, sampled at the rate n_h/N_h.

We start with the simplest auxiliary vector and proceed to more complex ones. We consider various specifications of the auxiliary vector $\mathbf{x}_k = \mathbf{x}_k^*$ in the InfoU

Estimation in Surveys with Nonresponse   C.-E. Särndal and S. Lundström
© 2005 John Wiley & Sons, Ltd

case and $\mathbf{x}_k = \mathbf{x}_k^\circ$ in the InfoS case. The InfoUS mixture case, which postulates information of both kinds, is considered in later chapters.

A calibration estimator for a domain total follows from the principle expressed by (6.12), so we need not further discuss domain estimation in this chapter.

7.2. THE SIMPLEST AUXILIARY VECTOR

The simplest auxiliary vector, whose value is constant for all k, has already been mentioned in Example 6.1. It recognizes no differences among elements and is ineffective for nonresponse treatment. For InfoU with $\mathbf{x}_k = \mathbf{x}_k^* = 1$ for all k, the calibration estimator (6.11) becomes

$$\hat{Y}_{\mathrm{W}} = N\bar{y}_{r;d} = \hat{Y}_{\mathrm{EXP}} \tag{7.1}$$

where $\bar{y}_{r;d} = \sum_r d_k y_k / \sum_r d_k$. The subscript EXP reflects the often-used term *(straight) expansion estimator*. In the literature the term is often reserved more specifically for the form that (7.1) takes under SI. In that case, with $d_k = N/n$ for all k, we have

$$\hat{Y}_{\mathrm{EXP}} = \frac{N}{m} \sum_r y_k = N\bar{y}_r. \tag{7.2}$$

The bias of $\hat{Y}_{\mathrm{EXP}}$ can be very large, because of the extremely weak auxiliary information. Exceptionally, $\hat{Y}_{\mathrm{EXP}}$ may still find use if there is no better auxiliary information and/or if there is strong reason to believe that the nonresponse occurs at random. Also, $\hat{Y}_{\mathrm{EXP}}$ is sometimes computed in a survey just to provide a benchmark estimate to which better alternatives can be compared.

Remark 7.1

The mean $\bar{y}_{r;d}$ exemplifies a system of notation that we use for various quantities. The overbar indicates a (weighted) mean. The general rule is that the first index denotes the set over which the quantity is defined or computed. If the first index is U, the quantity is nonrandom; if this index is r or s, the quantity is random. Thus, in the case of $\bar{y}_{r;d} = \sum_r d_k y_k / \sum_r d_k$, the computation is done over the elements of the response set r, using $d_k = 1/\pi_k$ as weights. The absence of the second index implies that the weighting is uniform, as in $\bar{y}_r = (1/m) \sum_r y_k$.   □

7.3. ONE-WAY CLASSIFICATION

Information about a classification of elements is highly useful for estimation with nonresponse. Consider a classification concept allowing an element k to be placed into one out of a set of P mutually exclusive and exhaustive categories or groups – for example, age groups, or the groups defined by the cross-classification age group by sex by occupational group. The group identifier for element k is defined as

$$\boldsymbol{\gamma}_k = (\gamma_{1k}, \ldots, \gamma_{pk}, \ldots, \gamma_{Pk})' \tag{7.3}$$

where, for $p = 1, \ldots, P, \gamma_{pk} = 1$ if k belongs to group p and $\gamma_{pk} = 0$ if not. That is, the vector $\boldsymbol{\gamma}_k$ has $P - 1$ zero components and one single unit component. The latter identifies the group membership of k. Information about $\boldsymbol{\gamma}_k$ for InfoU and information about $\boldsymbol{\gamma}_k$ for InfoS represent two different cases; we consider both. The condition $\boldsymbol{\mu}'\mathbf{x}_k = 1$ for all k holds by taking $\boldsymbol{\mu}' = (1, 1, \ldots, 1)$, so formula (6.15) applies.

In the InfoU case, the auxiliary vector is $\mathbf{x}_k = \mathbf{x}_k^* = \boldsymbol{\gamma}_k$ with the associated information $\sum_U \mathbf{x}_k^* = (N_1, \ldots, N_p, \ldots, N_P)'$, where N_p is the known size of the group U_p. This applies to many surveys, particularly in the Scandinavian countries, where available registers of the total population can generate a variety of classifications for individuals based on their demographic, educational, economic and other personal characteristics. The population frequencies N_p are established by a simple count in the register. In other countries, the N_p may consist of (updated) census counts. Let r_p be the response set from group p; the total response set is $r = \bigcup_{p=1}^{P} r_p$. From the standard weighting formula (6.10), or from (6.15) with $d_{\alpha k} = d_k$ and $\mathbf{z}_k = \mathbf{x}_k$, we obtain $v_k = F_p^*$ for all $k \in r_p$, where $F_p^* = N_p / \sum_{r_p} d_k$. The calibration estimator (6.11) becomes

$$\hat{Y}_{\mathrm{W}} = \sum_{p=1}^{P} N_p \overline{y}_{r_p;d} = \hat{Y}_{\mathrm{PWA}} \qquad (7.4)$$

where $\overline{y}_{r_p;d} = \sum_{r_p} d_k y_k / \sum_{r_p} d_k$ is the design-weighted group mean for respondents. If the nonresponse occurs at random within every group, then $\overline{y}_{r_p;d}$ estimates the group mean $\overline{y}_{U_p} = \sum_{U_p} y_k / N_p$ with essentially no bias, and then (7.4) is nearly without bias for $Y = \sum_U y_k$.

In particular, for SI and InfoU, (7.4) becomes

$$\hat{Y}_{\mathrm{PWA}} = \sum_{p=1}^{P} N_p \overline{y}_{r_p} \qquad (7.5)$$

where $\overline{y}_{r_p} = \sum_{r_p} y_k / m_p$ is the straight y-mean for the m_p respondents in group p. Formula (7.5) is sometimes called the *poststratified estimator*. The term is mildly misleading in that the traditional poststratified estimator refers to a single phase of sampling. Consequently, authors such as Kalton and Kasprzyk (1986) make a distinction between the poststratified estimator, as used for full response in single-phase sampling, and (7.5), which they call the *population weighting adjustment estimator*, as reflected in our notation $\hat{Y}_{\mathrm{PWA}}$. In the latter, more appropriate term lies a recognition of a sampling phase followed by a nonresponse phase. Over the years, a number of authors have discussed (7.5), including Bethlehem and Kersten (1985) and Thomsen (1973, 1978). Important as (7.5) may be, it is only one of the simplest applications of the much more general calibration method.

Now consider InfoS with the auxiliary vector $\mathbf{x}_k = \mathbf{x}_k^\circ = \boldsymbol{\gamma}_k$. In this case, the information at hand states that, for every $k \in s$ (and thus for every $k \in r$), we can place k into the group where it belongs. Let s_p, of size n_p, be the part of the sample s falling in group p; $s = \bigcup_{p=1}^{P} s_p$. The following are examples of situations in which a classification may be established for every $k \in s$ (but not necessarily for every $k \in U$):

(i)   Every sampled element can be attributed to one out of P possible interviewers responsible for the data collection.

(ii) The categories represent P different socioeconomic groups for which the population sizes are not available for use in the estimation.

(iii) In a telephone survey with repeated attempts to contact, the elements $k \in s$ are grouped by the attempt (first, second, ..., Pth) at which contact is established.

In (iii), telephone contact with an element leads either to participation in the survey or to refusal, hence nonresponse. Empirical evidence suggests that the participation rate m_p/n_p is higher among elements that can be reached early on, that is, for $p = 1$ or 2, than for, say, $p = 5$ or 6. The groups are thus related to response propensity. (For the final group, n_P is the number of elements in s having so far not been reached, although some may have been reached had the procedure continued.)

Although N_p is unknown, there is information enough to compute $\hat{N}_p = \sum_{s_p} d_k$ for $p = 1, \ldots, P$. From (6.10) we obtain $v_k = F_p^\circ$ for all $k \in r_p$, where $F_p^\circ = \hat{N}_p / \sum_{r_p} d_k$. The calibration estimator (6.11) becomes

$$\hat{Y}_{\mathrm{W}} = \sum_{p=1}^{P} \hat{N}_p \bar{y}_{r_p;d} = \hat{Y}_{\mathrm{WC}}. \tag{7.6}$$

In particular, consider SI of n from N, and InfoS. Out of the n sampled elements, let n_p be the number that happen to fall in group p, of which m_p respond. The weights obtained from (6.10) are then $w_k = Nn_p/nm_p$. They reflect an 'expansion by groups, using the inverse group response rate'. Then (7.6) becomes

$$\hat{Y}_{\mathrm{WC}} = \sum_{p=1}^{P} \hat{N}_p \bar{y}_{r_p} \tag{7.7}$$

with $\hat{N}_p = Nn_p/n$. Known as *the weighting class estimator*, and therefore denoted $\hat{Y}_{\mathrm{WC}}$, it has also been often discussed in earlier literature, as, for example, in Oh and Scheuren (1983), Kalton and Kasprzyk (1986), Little (1986) and Statistics Sweden (1980).

Although it may seem surprising at first, (7.4) and (7.6) have essentially the same nonresponse bias. This property will be seen in Chapter 10. The difference in bias is negligible, even though $\hat{Y}_{\mathrm{PWA}}$ uses information at a higher level than $\hat{Y}_{\mathrm{WC}}$. This shows up in the variance: (7.4) usually has the smaller variance of the two.

The STSI sampling design plays a prominent role in survey practice and deserves special attention. Consider STSI and InfoU, in such a way that each stratum is also taken as a group for calibration purposes. That is, the index p in the InfoU vector $\mathbf{x}_k = \mathbf{x}_k^* = \boldsymbol{\gamma}_k = (\gamma_{1k}, \ldots, \gamma_{pk}, \ldots, \gamma_{Pk})'$ indicates one out of P

possible strata. The sampling weights are $d_k = N_p/n_p$, the inverse of the sampling rate in stratum p. We have $v_k = F_p^* = n_p/m_p$ for all $k \in r_p$, leading to $w_k = d_k v_k = (N_p/n_p)(n_p/m_p) = N_p/m_p$. Then (7.4) becomes

$$\hat{Y}_W = \sum_{p=1}^{P} N_p \bar{y}_{r_p}. \tag{7.8}$$

Although identical in form to (7.5), (7.8) is conceptually different in that each group is also a stratum for sampling purposes. The calibration approach thus provides justification for a common practice under STSI, namely, of expanding the design weights N_p/n_p stratum by stratum, using the inverse of the stratum response rate n_p/m_p. For STSI and InfoS, we have $\mathbf{x}_k = \mathbf{x}_k^\circ = \mathbf{\gamma}_k$, and we again arrive at (7.8).

Remark 7.2

Some would argue that we should *first* adjust the weights stratum by stratum, and *then* apply a calibration. That is, *before* calibration we adjust the sampling weights within each stratum, using the inverse of the stratum response rate, n_p/m_p, so that the initial weights are $d_{ak} = (N_p/n_p)(n_p/m_p) = N_p/m_p$ for $k \in r_p$. This case is not covered by Remark 6.9 because we change the design weights stratum by stratum, and not by a factor that is constant throughout. Now use (6.14) with the initial weights $d_{ak} = N_p/m_p$ and $\mathbf{x}_k = \mathbf{x}_k^* = \mathbf{z}_k = \mathbf{\gamma}_k$. We obtain $v_k = 1$ for all k. The final weights are exactly the same as in (7.8), $w_k = N_p/m_p$ for $k \in r_p$. The calibration step changes nothing here; initial weights and final weights coincide.

However, it is not true more generally that the two alternatives for initial weights under STSI, $d_{ak} = d_k = N_p/n_p$ and $d_{ak} = N_p/m_p$, give identical final weights. It depends on the $\mathbf{x}_k$-vector. Section 8.2 gives an example of an $\mathbf{x}_k$-vector for which the two alternatives lead to different final weights w_k. The difference between the two calibrated estimates is likely to be inconsequential in most cases, but the question merits further study. □

The estimators (7.4), (7.6) and (7.8) can be computed by a number of existing software programs. For STSI, CLAN97 can also compute the weights and calibration estimators when a different set of groups is specified inside each stratum.

7.4. A SINGLE QUANTITATIVE AUXILIARY VARIABLE

This example is of a rare kind in that the objective of arriving at a well-known formula requires an instrument vector $\mathbf{z}_k$ other than the standard choice $\mathbf{z}_k = \mathbf{x}_k$. We consider a quantitative auxiliary variable, x, with value x_k for element k. It may be a measure of the size of k, such as the number of employees of enterprise k in a business survey. For every $k \in r$, the value x_k is observed or otherwise known. Two cases arise: we can work with x_k alone, or, supposing that N is known, with the vector $(1, x_k)'$. We consider both possibilities.

Consider InfoU with $\mathbf{x}_k = \mathbf{x}_k^* = x_k$ and the corresponding known population total $\sum_U x_k$. From (6.14) with $d_{\alpha k} = d_k$ and $\mathbf{z}_k = 1$ for all k, we obtain

$$\hat{Y}_W = \left(\sum_U x_k \right) \frac{\sum_r d_k y_k}{\sum_r d_k x_k} = \hat{Y}_{RA}. \qquad (7.9)$$

This has the *ratio estimator* form, hence the index RA. If the nonresponse occurs with equal probability throughout the population, then (7.9) is nearly without bias for $Y = \sum_U y_k$.

The ratio estimator feature becomes even more explicit under SI, in which case

$$\hat{Y}_{RA} = N \overline{x}_U \frac{\overline{y}_r}{\overline{x}_r} \qquad (7.10)$$

where $\overline{x}_U = \sum_U x_k / N$, $\overline{y}_r = \sum_r y_k / m$, and $\overline{x}_r$ is defined analogously. Note that the ratio estimator usually discussed in standard textbooks is (4.24), which is the complete response version of (7.9), obtained when $r = s$. Under SI, (4.24) is nearly unbiased, but such a property cannot be claimed for (7.10) unless the nonresponse occurs completely at random.

In the InfoS case, we know the value x_k for every $k \in s$. We obtain the weights from (6.14) with $\mathbf{x}_k = \mathbf{x}_k^\circ = x_k$, $d_{\alpha k} = d_k$ and $\mathbf{z}_k = 1$ for all k. In particular, for SI and InfoS, the sample x-mean $\overline{x}_s = \sum_s x_k / n$ is computable, and the estimator becomes

$$\hat{Y}_W = N \overline{x}_s \frac{\overline{y}_r}{\overline{x}_r}. \qquad (7.11)$$

In the presence of a continuous x-variable, we can alternatively formulate the auxiliary vector as $\mathbf{x}_k = (1, x_k)'$. For InfoU, we have $\mathbf{x}_k = \mathbf{x}_k^* = (1, x_k)'$ with the information input $\mathbf{X} = \sum_U \mathbf{x}_k = \left(N, \sum_U x_k \right)'$, assuming that the population size N is known. With the standard weighting (6.10) we obtain

$$\hat{Y}_W = N\{\overline{y}_{r;d} + (\overline{x}_U - \overline{x}_{r;d}) B_{r;d}\} = \hat{Y}_{REG} \qquad (7.12)$$

the *regression estimator* form, where $\overline{x}_{r;d} = \sum_r d_k x_k / \sum_r d_k$ and

$$B_{r;d} = \frac{\sum_r d_k (x_k - \overline{x}_{r;d})(y_k - \overline{y}_{r;d})}{\sum_r d_k (x_k - \overline{x}_{r;d})^2}. \qquad (7.13)$$

The merits of $\hat{Y}_{REG}$ as a nonresponse-adjusted estimator are discussed, for example, in Bethlehem (1988). The classical simple regression estimator, analysed in many textbooks for the case of complete response, is (4.25). It is obtained from (7.12) when $r = s$. Estimator (7.12) usually gives better protection against nonresponse bias than the ratio estimator (7.9).

Remark 7.3

The ratio estimator has a long history of excellent performance in survey theory, as long as there is no nonresponse. To uncritically extend the use of the ratio estimator to surveys with nonresponse is risky. One cannot strongly recommend

(7.9) or (7.10). For these to be nearly unbiased, the response probabilities must satisfy a very strong condition: they need to be equal throughout the population, for reasons shown in Chapter 10. The ratio estimator is thus handicapped, compared to alternatives such as the one-way classification estimator discussed in Section 7.3. The latter is nearly unbiased under equal response probability within groups, something which is easier to satisfy than equal response probability throughout the whole population. This observation points to a modified and more reliable use of a quantitative x-variable: form P groups defined by nonoverlapping intervals of the values x_k available for $k \in U$ or for $k \in s$; then the $\mathbf{x}_k$-vector becomes a classification vector $\boldsymbol{\gamma}_k = (\gamma_{1k}, \ldots, \gamma_{pk}, \ldots, \gamma_{Pk})'$. Alternatively one could formulate $\mathbf{x}_k$ in the manner of (7.16) in the following section. □

7.5. ONE-WAY CLASSIFICATION COMBINED WITH A QUANTITATIVE VARIABLE

In this application, auxiliary information is present both about a P-valued categorical variable and about a quantitative variable, x. The latter may measure the size of an element. Our information about a responding element k consists of the observed or otherwise known value x_k and the group membership, coded by the group indicator $\boldsymbol{\gamma}_k = (\gamma_{1k}, \ldots, \gamma_{pk}, \ldots, \gamma_{Pk})'$ defined as in Section 7.3. Again two cases arise. To form the auxiliary vector, we can work with x_k by itself, or alternatively with the vector $(1, x_k)'$. We start with the first possibility.

For InfoU, we have $\mathbf{x}_k = \mathbf{x}_k^* = (\gamma_{1k}x_k, \ldots, \gamma_{pk}x_k, \ldots, \gamma_{Pk}x_k)'$. The required information input, $\sum_U \mathbf{x}_k^*$, is the vector composed of the P group totals $\sum_{U_p} x_k$. A formulation of $\mathbf{x}_k$ that also capitalizes on known group sizes N_p is considered later in this section. However, from (6.14) with $d_{\alpha k} = d_k$ and $\mathbf{z}_k = \boldsymbol{\gamma}_k = (\gamma_{1k}, \ldots, \gamma_{pk}, \ldots, \gamma_{Pk})'$ we obtain another well-known estimator form

$$\hat{Y}_W = \sum_{p=1}^{P} \left(\sum_{U_p} x_k \right) \frac{\overline{y}_{r_p;d}}{\overline{x}_{r_p;d}} = \hat{Y}_{\text{SEPRA}} \tag{7.14}$$

where $\overline{y}_{r_p;d}$ was defined in Section 7.3 and $\overline{x}_{r_p;d}$ is analogous. We recognize $\hat{Y}_{\text{SEPRA}}$ as a *separate ratio estimator*, constructed as a sum of ratio estimators, one for each group.

For InfoS, the auxiliary vector is $\mathbf{x}_k = \mathbf{x}_k^\circ = (\gamma_{1k}x_k, \ldots, \gamma_{pk}x_k, \ldots, \gamma_{Pk}x_k)'$. From (6.14) with $d_{\alpha k} = d_k = N/n$ and $\mathbf{z}_k = \boldsymbol{\gamma}_k$, we obtain $w_k = \sum_{s_p} d_k x_k / \sum_{r_p} d_k x_k$ for all k in group p. The result is another easily recognizable estimator,

$$\hat{Y}_W = \sum_{p=1}^{P} \left(\sum_{s_p} d_k x_k \right) \frac{\overline{y}_{r_p;d}}{\overline{x}_{r_p;d}}. \tag{7.15}$$

The telling description 'weighting by groups using a size variable' is sometimes used, as in Statistics Sweden (1980). Compared to $\hat{Y}_{\text{RA}}$ given by (7.9), both (7.14) and (7.15) offer improved protection against nonresponse bias, as will seen more clearly later.

To incorporate information that may exist about the population group sizes N_p, we formulate instead the auxiliary vector for InfoU as

$$\mathbf{x}_k = \mathbf{x}_k^* = (\gamma_{1k}, \ldots, \gamma_{pk}, \ldots, \gamma_{Pk}, \gamma_{1k}x_k, \ldots, \gamma_{pk}x_k, \ldots, \gamma_{Pk}x_k)'. \quad (7.16)$$

With weights obtained by (6.10), the estimator is

$$\hat{Y}_\mathrm{W} = \sum_{p=1}^{P} N_p \{\bar{y}_{r_p;d} + (\bar{x}_{U_p} - \bar{x}_{r_p;d})B_{r_p;d}\} = \hat{Y}_\mathrm{SEPREG} \quad (7.17)$$

where $\bar{x}_{U_p} = \sum_{U_p} x_k / N_p$, $\bar{y}_{r_p;d}$ and $\bar{x}_{r_p;d}$ are as already defined, and $B_{r_p;d}$ is given by (7.13) if we replace r by r_p. The notation $\hat{Y}_\mathrm{SEPREG}$ suggests another well-known form, that of the *separate regression estimator*. Of the estimators seen so far in this chapter, $\hat{Y}_\mathrm{SEPREG}$ has the best chance of effective protection against nonresponse bias, for reasons discussed in Chapter 10.

7.6. TWO-WAY CLASSIFICATION

In practice information often exists on two or more categorical auxiliary variables or factors. Here we consider the case of two factors. The reasoning can easily be extended to a multi-factor classification.

For the first factor, suppose as in Section 7.3 that there are P categories indexed $p = 1, \ldots, P$, representing, for example, a geographical classification. For the second factor, let there be H categories indexed $h = 1, \ldots, H$, representing for example a socioeconomic classification. A responding element k can be placed into one of the $P \times H$ cells formed by the cross-classification of the two factors. If the procedure in Section 7.3 is followed, we formulate a one-way classification auxiliary vector with dimension $P \times H$. For InfoU, this requires that the $P \times H$ population cell counts N_{ph} are known. But other, more parsimonious formulations or the auxiliary vector are possible or perhaps necessary.

Consider the indicator vector

$$(\gamma_{1k}, \ldots, \gamma_{pk}, \ldots, \gamma_{Pk}, \delta_{1k}, \ldots, \delta_{hk}, \ldots, \delta_{Hk})' \quad (7.18)$$

where the γs indicate the first classification with P groups and the δs indicate the second classification with H groups. That is, γ_{pk} is as in Section 7.3, while $\delta_{hk} = 1$ if k is a member of group h, and $\delta_{hk} = 0$ if not, for $h = 1, \ldots, H$.

We must formulate the auxiliary vector $\mathbf{x}_k$ so as to avoid a singularity of the matrix that needs to be inverted when the calibrated weights are computed. There is a linear dependence in the vector (7.18): for every k we have $\sum_{p=1}^{P} \gamma_{pk} = \sum_{h=1}^{H} \delta_{hk} = 1$, and the matrix to invert is singular. Let us, therefore, drop any one of the δs in the second classification, say, the last. For InfoU, the auxiliary vector is

$$\mathbf{x}_k = \mathbf{x}_k^* = (\gamma_{1k}, \ldots, \gamma_{pk}, \ldots, \gamma_{Pk}, \delta_{1k}, \ldots, \delta_{hk}, \ldots, \delta_{H-1,k})'. \quad (7.19)$$

Its dimension is $P + H - 1$. The corresponding information input, $\sum_U \mathbf{x}_k = \sum_U \mathbf{x}_k^*$, is a vector consisting of the $P + H - 1$ marginal counts, $N_{p\bullet} = \sum_{h=1}^{H} N_{ph}, p = 1, \ldots, P$, and $N_{\bullet h} = \sum_{p=1}^{P} N_{ph}, h = 1, \ldots, H - 1$, where N_{ph}

is the population count in cell ph. We thus obtain the procedure known as 'calibration on marginal counts', mentioned earlier in Example 4.5. The auxiliary vector formulation (7.19) allows us to treat three commonly occurring situations:

(i)   The $P \times H$ cell counts, $N_{ph}, p = 1, \ldots, P, h = 1, \ldots, H$, are known, but it is considered that the marginal counts, $N_{p\bullet}, p = 1, \ldots, P$, and $N_{\bullet h}, h = 1, \ldots, H - 1$, contain almost as much information, so we instead use (7.19), incurring no serious loss.

(ii) The $P \times H$ cell counts, $N_{ph}, p = 1, \ldots, P, h = 1, \ldots, H$, are known, but some of them are extremely small or zero, a situation frequently arising in practice. Collapsing of cells is an alternative remedy for this problem.

(iii) The marginal counts, the $N_{p\bullet}$ and the $N_{\bullet h}$, are known, but not the cell counts N_{ph}. An important instance of this occurs when the $N_{p\bullet}$ and the $N_{\bullet h}$ are counts derived from two different registers. They need not be matched on an element basis, which can be a considerable advantage. Two possibilities exist for preparing the InfoU scenario: (a) to obtain class memberships for $k \in r$, the response set is matched, using a unique key such as the personal identification number, with the first register, then with the second, while $N_{p\bullet}$ and $N_{\bullet h}$ are established by counting in each register separately; (b) the class membership information is solicited in the survey and obtained for every $k \in r$, while the counts are imported, $N_{p\bullet}$ for $p = 1, \ldots, P$ from one register and $N_{\bullet h}$, for $h = 1, \ldots, H$, from the other.

For the auxiliary vector (7.19), the standard weighting w_k obtained by (6.10) gives the calibration estimator

$$\hat{Y}_{\mathrm{W}} = \sum_r w_k y_k = \hat{Y}_{\mathrm{TWOWAY}}. \tag{7.20}$$

The weights w_k do not have any easily presentable form, but there are no computational obstacles to obtaining them. The weights and the estimator $\hat{Y}_{\mathrm{TWOWAY}}$ are readily computed using existing software such as CLAN97 or CALMAR. The latter is described in Deville et $al.$ (1993).

The theoretical underpinnings of calibration for cross-classifications are given, for example, in Deville and Särndal (1992), where the emphasis is on the case of complete response. Extensions to nonresponse are discussed in Deville (2000) and Caron et $al.$ (1998), the latter containing a description of the French software POULPE.

Another familiar name in connection with information on marginal counts is the $raking$ $ratio$ $method$, discussed, for example, in Oh and Scheuren (1983). The computational algorithm used in the raking ratio method delivers a point estimator that is not exactly identical to $\hat{Y}_{\mathrm{TWOWAY}}$, as described in this section, but usually the numerical difference between the two is inconsequential.

7.7. A MONTE CARLO SIMULATION STUDY

The objective in this section is to give an empirical illustration of the calibration estimators derived in Sections 7.2–7.6. We wish to see how the different inputs of information that they represent become reflected primarily in the bias, and secondarily in the variance of the estimator.

To this end we created a population from a real data set derived from the 1992 survey of the 965 Swedish clerical municipalities. We refer to the population created as KYBOK, which is also the acronym for the survey. The KYBOK population is used in several empirical illustrations in this book. For the 965 clerical municipalities, data are available from administrative sources on a number of variables, such as revenues, costs, expenses and investments. Furthermore, the municipalities can be classified into four groups according to managerial type. Data on other variables are solicited by a postal inquiry to which 832 municipalities responded. The 1992 response rate was thus 86%.

We defined the population U for this experiment to consist of the $N = 832$ responding elements. Expenditure on administration and maintenance was chosen as the study variable y_k. We realized repeated samples and response sets from this population of size 832.

An auxiliary variable for this experiment was a numerical variable whose value x_k for element k is defined as the square root of revenue advances. (The square root mitigates the influence of very large values of revenue advances for the largest elements.)

Table 7.1 shows some key characteristics of the population. It is clear that the grouping from 1 to 4 reflects a progression from smaller to larger clerical municipalities, and those in group 4 are comparatively few.

In a real survey, the response distribution is unknown. Therefore a simulation should attempt to cover a range of plausible response distributions. This would allow us to see how an estimator behaves under different conditions. For practical reasons, this is seldom possible. Here we restrict ourselves to two response distributions. They were created as follows.

The first, called the *logit response distribution*, has a background in reality. It is based on the fit of a logistic model to all 965 elements. The dependent variable for this fit is the dichotomous variable 'responding/not responding'. The predictor variables used in this fit were revenue advances for element k, and two other numerical predictors for which data are also available for all 965 elements. The grouping into four types of clerical municipality was ignored in this model fit.

Table 7.1   Key characteristics of the KYBOK study population.

Characteristic	Overall	By group			
		1	2	3	4
Number of elements	832	218	272	290	52
Total of y	1 025 983	132 788	241 836	343 359	308 000
Mean of y	1 233	609	889	1 184	5 923
Variance of $y \times 10^{-3}$	3 598	228	566	1 683	20 369
Coefficient of correlation between y and x	0.92	0.87	0.90	0.84	0.91

The average of the estimated response probabilities is 86 %. The resulting fitted values, all belonging to the unit interval $(0,1)$, are used as response probabilities θ_k in creating the response sets in the simulation.

The second response distribution, called the *increasing exponential response distribution*, is a pure construction. Its response probabilities were computed as $\theta_k = 1 - \exp(-c_1 x_k)$, where $c_1 = 0.0318$ was used to obtain the desired average response probability of 86 %.

Over a sequence of repeated samples, the nonresponse rate achieved with the aid of these two response distributions will thus vary around 14 %. This rate is rather modest, compared with many real-life surveys. Nevertheless, we shall find that the harmful effects, mainly the bias, of such a nonresponse rate are severe, especially for estimators that do not sufficiently engage auxiliary information to counteract the tendency to bias.

For logit response distribution, the response probability θ_k has a tendency to increase with x_k. That the response probability increases with x_k is evident for the increasing exponential response distribution. The mean response probability by group is not very much different for the two response distributions. The variability of the θ_k is somewhat more pronounced for the increasing exponential distribution. Despite a resemblance between the two response distributions, we find that they produce remarkably different effects on a number of issues in the study.

We drew 10 000 samples each of size $n = 300$ by SI from the population U of size 832. For each sample, two response sets (one for each response distribution) were created by carrying out, for the sampled element k, a Bernoulli experiment with parameter θ_k. That is, the element k becomes a respondent with probability θ_k and a non-respondent with probability $1 - \theta_k$, where θ_k is known for all k in both response distributions. For every sample s, and for each response distribution, 300 independent Bernoulli experiments were thus carried out to obtain the response set r. In the end we have 10 000 response sets r_j, $j = 1, \ldots, 10\,000$, for each response distribution. For every response set, we computed eight different estimators.

The auxiliary information used, in its entirety or in part, to construct the estimators, was: (i) the value of a one-way classification vector of dimension 4, denoted γ_k and indicating the type of clerical municipality; and (ii) the value x_k, defined as the square root of revenue advances. The eight estimators studied in the experiment are the SI versions of formulae (7.1), (7.4), (7.6), (7.9), (7.12), (7.14), (7.17) and (7.20). We refer to these by their respective abbreviated names EXP, PWA, WC, RA, REG, SEPRA, SEPREG and TWOWAY.

For TWOWAY, the first grouping is by the $P = 4$ types of clerical municipality, and the second grouping with $H = 4$ groups was created using a method suggested by Wright (1983). We size-ordered the 832 elements from the smallest to the largest according to the value of x_k. The first group consists of the first elements of the size ordering, up to a point where the sum of the x-values equals (approximately) $\sum_U x_k/4$. The second group consists of the elements accounting for the next quarter of $\sum_U x_k$, and so on. This procedure resulted in four groups with sizes 348, 234, 161 and 89, respectively. The resulting vector $\mathbf{x}_k$, given by (7.19) with $P = H = 4$, has dimension $4 + 4 - 1 = 7$. Compared to using x_k in its continuous form, the second grouping causes some loss of information, but, as argued elsewhere in this book, in the presence of nonresponse it is sometimes preferable to transform a continuous variable into a size-grouped one.

Let $\hat{Y}_W$ be one of the estimators participating in the experiment, and let $\hat{Y}_{W(j)}$ be the value computed for this estimator for the response set r_j, $j = 1, \ldots, 10\,000$. We computed two performance measures: (i) the simulation relative bias,

$$RB_{\text{SIM}}(\hat{Y}_W) = \frac{E_{\text{SIM}}(\hat{Y}_W) - Y}{Y}$$

where

$$E_{\text{SIM}}(\hat{Y}_W) = \sum_{j=1}^{10\,000} \hat{Y}_{W(j)}/10\,000$$

is the simulation expectation of $\hat{Y}_W$; and (ii) the simulation variance,

$$V_{\text{SIM}}(\hat{Y}_W) = \frac{1}{9999} \sum_{j=1}^{10\,000} [\hat{Y}_{W(j)} - E_{\text{SIM}}(\hat{Y}_W)]^2.$$

The results are displayed in Table 7.2.

Table 7.2 Simulation relative bias of $\hat{Y}_W$ (%) and simulation variance of $\hat{Y}_W$ (multiplied by $\times 10^{-8}$) for two response distributions and different point estimators. Samples of size $n = 300$ drawn by SI. Nonresponse rate 14 %.

Response distribution	Estimator	$RB_{\text{SIM}}(\hat{Y}_W)$ $\times 10^2$	$V_{\text{SIM}}(\hat{Y}_W)$ $\times 10^{-8}$
Logit	Expansion (EXP)	5.0	69.6
	Weighting class (WC)	2.2	59.4
	Population weighting adjustment (PWA)	2.2	37.1
	Ratio (RA)	2.5	27.5
	Separate ratio (SEPRA)	0.7	11.8
	Regression (REG)	−0.6	9.5
	Separate regression (SEPREG)	−0.2	8.1
	Two-way classification (TWOWAY)	0.5	21.7
Increasing exponential	Expansion (EXP)	9.3	70.1
	Weighting class (WC)	5.7	57.7
	Population weighting adjustment (PWA)	5.7	36.3
	Ratio (RA)	3.9	26.1
	Separate ratio (SEPRA)	2.0	11.3
	Regression (REG)	−2.7	8.1
	Separate regression (SEPREG)	−0.8	7.4
	Two-way classification (TWOWAY)	0.5	20.3

The eight estimators compare very similarly for the two response distributions. We can discuss results for both at the same time.

The estimator EXP uses the known population size N as the only input of auxiliary information. No effective adjustment for nonresponse is achieved. As expected, the bias of EXP is large, comparatively speaking, and so is its variance.

The most complete use of the available information is present in SEPREG. Not surprisingly, SEPREG gives good results. Its absolute bias is the lowest for increasing exponential, and second lowest for logit. Its variance is the lowest in both cases.

The other estimators fall in between the two extremes, EXP and SEPREG, in their use of the available information. This is reflected in both the bias and the variance. The use of the one-way classification defined by γ_k gives a considerable improvement, evidenced by the considerably lower bias of WC and PWA, as compared to EXP. That the classification has a favourable effect on bias and on variance is also seen by comparing RA with SEPRA, and REG with SEPREG.

TWOWAY has a small bias, but its variance exceeds that of SEPREG, because making size groups out of the x_k-values causes some loss of information, compared to using the x_k as continuous measurements.

It is no coincidence that WC and PWA give the same simulation bias, to the number of decimal places reported in the table. Section 9.1 gives the theoretical explanation for this equality. The variance is much greater for WC than for PWA, since PWA uses the population auxiliary counts, whereas WC estimates these counts.

Information about the group sizes N_p distinguishes SEPREG from SEPRA. The difference may seem minor. Nevertheless SEPREG performs considerably better, with both a smaller bias and a smaller variance.

Remark 7.4

The relative bias of a calibration estimator $\hat{Y}_W$ can also be evaluated by a formula, computable when the whole population is specified, as it is in this simulation experiment. The bias of $\hat{Y}_W$ is given, to close approximation, by formula (9.14) in Section 9.4, so the relative bias of $\hat{Y}_W$ is approximately $-\sum_U (1 - \theta_k)e_{\theta k}/Y$, where $e_{\theta k}$ is explained in Proposition 9.1. A computation for the KYBOK population of the quantity $-\sum_U (1 - \theta_k)e_{\theta k}/Y$ gave values in very close agreement with the simulated relative bias, $RB_{SIM}(\hat{Y}_W)$, for all of the estimators $\hat{Y}_W$ in Table 7.2. The simulation thus supports the contention that (9.14) is a good approximation. $\square$

CHAPTER 8

The Combined Use of Sample Information and Population Information

8.1. OPTIONS FOR THE COMBINED USE OF INFORMATION

We turn now to the InfoUS case, where auxiliary information exists at both the sample level and the population level. The objective is to benefit from the combined information through a calibration that is as efficient as possible. The primary objective is a reduction of the nonresponse bias. A secondary objective is to achieve a small variance for the calibration estimator.

As in Section 6.2, the information at the sample level is carried by the moon vector, $\mathbf{x}_k^\circ$, whose value is known for every $k \in s$ and thus for every $k \in r$. The population-level information is carried by the star vector $\mathbf{x}_k^*$, for which we know the population total $\mathbf{X}^* = \sum_U \mathbf{x}_k^*$ and the individual value $\mathbf{x}_k^*$ for every $k \in r$.

A point emphasized in this chapter is that this combined information allows more than one alternative for computing calibrated weights, while still benefiting, in all the alternatives, from the total information. Estevao and Särndal (2002) show, in the context of two-phase sampling, that different options exist for the use of a total auxiliary information originating from two different levels, a population level and a sample level.

We can thus obtain different calibration estimators by using the information in different ways. Here we distinguish two types of procedure, single-step and two-step. Single-step is simpler and more direct; its auxiliary vector is the one given for InfoUS in Table 6.1. Two-step procedures, of which we consider two variations, two-step A and two-step B, have two computational steps, but usually give no reward in the form of a reduction of either the bias or the variance. We consider two-step A and two-step B because both, and the latter in particular, have found application in practice.

For the *single-step* procedure, the combined auxiliary vector is $\mathbf{x}_k = \begin{pmatrix} \mathbf{x}_k^* \\ \mathbf{x}_k^\circ \end{pmatrix}$.

The corresponding information input is $\mathbf{X} = \begin{pmatrix} \mathbf{X}^* \\ \hat{\mathbf{X}}^\circ \end{pmatrix}$, where $\mathbf{X}^* = \sum_U \mathbf{x}_k^*$ and

---

Estimation in Surveys with Nonresponse   C.-E. Särndal and S. Lundström
© 2005 John Wiley & Sons, Ltd

$\hat{\mathbf{X}}^\circ = \sum_s d_k \mathbf{x}_k^\circ$. Using the standard weighting (6.10), we have

$$w_k = d_k v_k, \qquad v_k = 1 + \boldsymbol{\lambda}_r' \mathbf{x}_k \tag{8.1}$$

where $\boldsymbol{\lambda}_r' = \left(\mathbf{X} - \sum_r d_k \mathbf{x}_k\right)' \left(\sum_r d_k \mathbf{x}_k \mathbf{x}_k'\right)^{-1}$. The resulting calibration estimator is $\hat{Y}_W = \sum_r w_k y_k$. It is sufficient for this chapter to consider the standard weighting (6.10). More generally, we could use weights of the type (6.14).

When information exists at two different levels, one can argue that the calibration could or should be carried out in two steps, accounting for one level at a time. This is a valid argument. We first calibrate from r to s on the information about $\mathbf{x}_k^\circ$ to obtain *intermediate weights* and then use these as initial weights in a second calibration from r to U, using the information about both $\mathbf{x}_k^*$ and $\mathbf{x}_k^\circ$. We call this procedure two-step A. A variation is to calibrate only on $\mathbf{x}_k^*$ in the second step, since the first step has already taken $\mathbf{x}_k^\circ$ into account. We call this procedure two-step B. Recall that each calibration step requires a set of initial weights and an auxiliary vector with a specified information input.

Now consider *procedure two-step A*. In step 1, use $d_k = 1/\pi_k$ as initial weights and compute intermediate weights w_k° by a calibration from r to s to satisfy the equation $\sum_r w_k^\circ \mathbf{x}_k^\circ = \sum_s d_k \mathbf{x}_k^\circ$. From (6.10) with $\mathbf{x}_k = \mathbf{x}_k^\circ$, we obtain the intermediate weights

$$w_k^\circ = d_k v_k^\circ, \qquad v_k^\circ = 1 + (\boldsymbol{\lambda}_r^\circ)' \mathbf{x}_k^\circ \tag{8.2}$$

with $(\boldsymbol{\lambda}_r^\circ)' = \left(\sum_s d_k \mathbf{x}_k^\circ - \sum_r d_k \mathbf{x}_k^\circ\right)' \left(\sum_r d_k \mathbf{x}_k^\circ (\mathbf{x}_k^\circ)'\right)^{-1}$. This step expands the insufficiently large sampling weights d_k to an appropriate level. In step 2, use the w_k° as initial weights, and produce the final weights by a calibration from r to U to satisfy $\sum_r w_k \mathbf{x}_k = \mathbf{X}$, where $\mathbf{x}_k = \begin{pmatrix} \mathbf{x}_k^* \\ \mathbf{x}_k^\circ \end{pmatrix}$ and $\mathbf{X} = \begin{pmatrix} \mathbf{X}^* \\ \hat{\mathbf{X}}^\circ \end{pmatrix}$. Using (6.14) with $d_{\alpha k} = w_k^\circ$ and the standard specification $\mathbf{z}_k = \mathbf{x}_k$, we obtain the final weights

$$w_{2Ak} = w_k^\circ v_k, \qquad v_k = 1 + \boldsymbol{\lambda}_r' \mathbf{x}_k \tag{8.3}$$

with $\boldsymbol{\lambda}_r' = \left(\mathbf{X} - \sum_r w_k^\circ \mathbf{x}_k\right)' \left(\sum_r w_k^\circ \mathbf{x}_k \mathbf{x}_k'\right)^{-1}$. The two-step A calibration estimator is $\hat{Y}_{W2A} = \sum_r w_{2Ak} y_k$.

The weights w_{2Ak} are correctly calibrated to the population information: $\sum_r w_{2Ak} \mathbf{x}_k^* = \sum_U \mathbf{x}_k^*$. Because $\mathbf{x}_k$ contains the moon vector $\mathbf{x}_k^\circ$, they also satisfy $\sum_r w_{2Ak} \mathbf{x}_k^\circ = \sum_s d_k \mathbf{x}_k^\circ$, just like the final weights in the single-step procedure. Note that the bottom part of the column vector $\left(\mathbf{X} - \sum_r w_k^\circ \mathbf{x}_k\right)$ in $\boldsymbol{\lambda}_r'$ is a vector of zeros, because

$$\left(\mathbf{X} - \sum_r w_k^\circ \mathbf{x}_k\right) = \begin{pmatrix} \mathbf{X}^* - \sum_r w_k^\circ \mathbf{x}_k^* \\ \mathbf{0} \end{pmatrix}$$

as a result of the first step.

The first step of *procedure two-step B* is the same as in two-step A. The intermediate weights are $w_k^\circ = d_k v_k^\circ$ with v_k° given by (8.2). Step 2 is also as in two-step A, except that $\mathbf{x}_k = \begin{pmatrix} \mathbf{x}_k^* \\ \mathbf{x}_k^\circ \end{pmatrix}$ and $\mathbf{X} = \begin{pmatrix} \mathbf{X}^* \\ \hat{\mathbf{X}}^\circ \end{pmatrix}$ are replaced with $\mathbf{x}_k = \mathbf{x}_k^*$

and $\mathbf{X} = \mathbf{X}^*$, respectively. The resulting final weights are

$$w_{2Bk} = w_k^{\circ} v_k, \qquad v_k = 1 + \boldsymbol{\lambda}_r' \mathbf{x}_k^* \qquad (8.4)$$

with $\boldsymbol{\lambda}_r' = \left(\sum_U \mathbf{x}_k^* - \sum_r w_k^{\circ} \mathbf{x}_k^* \right)' \left(\sum_r w_k^{\circ} \mathbf{x}_k^* (\mathbf{x}_k^*)' \right)^{-1}$. The two-step B calibration estimator is $\hat{Y}_{\mathrm{W2B}} = \sum_r w_{2Bk} y_k$.

The weight systems (8.1), (8.3) and (8.4) result from different computational procedures and will not in general be identical. Although the resulting three calibration estimators are not identical, all of them use the same information input. Therefore one can expect the estimators to have only minor differences, both with regard to the formulae and in their capacity to provide effective protection against nonresponse bias. We illustrate this analytically in Section 8.2 and empirically in Section 8.3.

The single-step procedure combines the two parts of the auxiliary information in a single, direct computation of final weights. There is no disadvantage in mixing the population information with the sample information. It is important that both sources of information are allowed to contribute. Two-step A and two-step B also use both sources. But the final weights in two-step B are less controlled, in that they fail to satisfy $\sum_r w_{2Bk} \mathbf{x}_k^{\circ} = \sum_s d_k \mathbf{x}_k^{\circ}$, which may have some undesirable effect on the bias and the variance of $\hat{Y}_{\mathrm{W2B}}$. A common feature of the three estimators is that they are derived without any reference to models or model fitting.

8.2. AN EXAMPLE OF CALIBRATION WITH INFORMATION AT BOTH LEVELS

We illustrate the different calibration procedures in Section 8.1 with an example in which the auxiliary information is sufficiently simple to render easily interpreted expressions. This allows us to see how the calibration estimator $\hat{Y}_{\mathrm{W}}$ changes with the choice of calibration procedure. We follow up, in Section 8.3, with an empirical comparison, based on a Monte Carlo simulation.

The information at the sample level concerns a one-way classification of elements as in Section 7.3. The group identifier for element k is

$$\boldsymbol{\gamma}_k = (\gamma_{1k}, \dots, \gamma_{pk}, \dots, \gamma_{Pk})' \qquad (8.5)$$

where $\gamma_{pk} = 1$ if k is a member of group p and $\gamma_{pk} = 0$ if not. The available information allows every $k \in s$ (and every $k \in r$) to be placed in the group to which it belongs. The moon vector is thus $\mathbf{x}_k^{\circ} = \boldsymbol{\gamma}_k$ for $k \in s$. The information at the population level concerns a continuous variable x with value x_k available for every element $k \in r$. The star vector is $\mathbf{x}_k^* = (1, x_k)'$ with the known population total $\sum_U \mathbf{x}_k^* = (N, X)'$, where $X = \sum_U x_k = N \bar{x}_U$. We refer to this combined information as WC/REG. It is the information in the estimator $\hat{Y}_{\mathrm{WC}}$ combined with that in the estimator $\hat{Y}_{\mathrm{REG}}$, as discussed in Sections 7.3 and 7.4.

We can compute $\hat{N}_p = \sum_s d_k \gamma_{pk} = \sum_{s_p} d_k$, where $s_p = s \cap U_p$, of size n_p, is the part of the sample s that belongs to group p. Let r_p, of size m_p, be the part of the response set r that belongs to group p. We omit the derivation of the weights w_k for the single-step, two-step A and two-step B procedures and state only the expression for the resulting calibration estimators. All three estimators contain a group expansion factor given by

$$F_p^\circ = \frac{\hat{N}_p}{\sum_{r_p} d_k} = \frac{\sum_{s_p} d_k}{\sum_{r_p} d_k}. \tag{8.6}$$

For some designs (8.6) simplifies to $F_p^\circ = n_p/m_p$, the inverse of the group response rate. This happens for SI, where the sampling weights are $d_k = N/n$, and for STSI such that every stratum coincides with one of the groups, in which case the sampling weights are $d_k = N_p/n_p$ for all k in stratum p of size N_p.

In the single-step procedure, we form $\hat{Y}_W = \sum_r w_k y_k$, using the weights (8.1) with $\mathbf{x}_k = \begin{pmatrix} x_k \\ \boldsymbol{\gamma}_k \end{pmatrix} = (x_k, \gamma_{1k}, \ldots, \gamma_{pk}, \ldots, \gamma_{Pk})'$. The information input is $\mathbf{X} = \begin{pmatrix} \sum_U x_k \\ \sum_s d_k \boldsymbol{\gamma}_k \end{pmatrix} = (X, \hat{N}_1, \ldots, \hat{N}_p, \ldots, \hat{N}_P)'$. Note that although $\mathbf{x}_k^* = (1, x_k)'$, the constant 1 is deleted from $\mathbf{x}_k$ to avoid a singularity that would otherwise affect the matrix requiring inversion. This causes no loss of information. We obtain the calibration estimator

$$\hat{Y}_W = \hat{Y}_{\text{SS:WC/REG}} = \sum_{p=1}^{P} F_p^\circ \left(\sum_{r_p} d_k(y_k - B_{\text{SSr};d} x_k) \right) + N \overline{x}_U B_{\text{SSr};d} \tag{8.7}$$

with

$$B_{\text{SSr};d} = \frac{\displaystyle\sum_{p=1}^{P} \sum_{r_p} d_k(x_k - \overline{x}_{r_p;d})(y_k - \overline{y}_{r_p;d})}{\displaystyle\sum_{p=1}^{P} \sum_{r_p} d_k(x_k - \overline{x}_{r_p;d})^2} \tag{8.8}$$

where $\overline{y}_{r_p;d} = \sum_{r_p} d_k y_k / \sum_{r_p} d_k$, and $\overline{x}_{r_p;d}$ is defined analogously. By Remark 6.9, the same final weights w_k are obtained if we initially apply a constant adjustment to all sampling weights. For example, the initial weights $d_{ak} = (n/m)d_k$, where m/n is the overall response rate, would lead to the same final weights.

We now turn to the two-step A procedure. In step 1, take $d_{ak} = d_k = 1/\pi_k$ as initial weights and use the auxiliary vector $\mathbf{x}_k = \mathbf{x}_k^\circ = \boldsymbol{\gamma}_k$, for which the information input is $\sum_s d_k \mathbf{x}_k^\circ = \sum_s d_k \boldsymbol{\gamma}_k = (\hat{N}_1, \ldots, \hat{N}_p, \ldots, \hat{N}_P)'$ with $\hat{N}_p = \sum_{s_p} d_k$. We obtain from (8.2) the intermediate weights $w_k^\circ = d_k F_p^\circ$ for $k \in r_p$. In step 2, take these w_k° as initial weights, and use the auxiliary vector $\mathbf{x}_k = (x_k, \boldsymbol{\gamma}_k')' = (x_k, \gamma_{1k}, \ldots, \gamma_{pk}, \ldots, \gamma_{Pk})'$ with $\mathbf{X} = (X, \hat{N}_1, \ldots, \hat{N}_p, \ldots, \hat{N}_P)'$. This information input is the same as in the single-step procedure, but the initial weights are $d_k F_p^\circ$ instead of d_k. The constant 1 must be deleted as in single-step. The final weights w_k, computed by (8.3), give the calibration estimator

$$\hat{Y}_{\text{2A:WC/REG}} = \sum_{p=1}^{P} F_p^\circ \left(\sum_{r_p} d_k(y_k - B_{\text{2Ar};d} x_k) \right) + N \overline{x}_U B_{\text{2Ar};d} \tag{8.9}$$

with

$$B_{2Ar;d} = \frac{\sum\limits_{p=1}^{P} F_p^{\circ} \sum\limits_{r_p} d_k (x_k - \bar{x}_{r_p;d})(y_k - \bar{y}_{r_p;d})}{\sum\limits_{p=1}^{P} F_p^{\circ} \sum\limits_{r_p} d_k (x_k - \bar{x}_{r_p;d})^2} \qquad (8.10)$$

where $\bar{y}_{r_p;d}$ and $\bar{x}_{r_p;d}$ are the same weighted group means as in single-step.

Finally, consider the two-step B procedure. The first step is as in two-step A. The intermediate weights $w_k^{\circ} = d_k F_p^{\circ}$ for $k \in r_p$ become initial weights for step 2, where the auxiliary vector in (8.4) is $\mathbf{x}_k^* = (1, x_k)'$ with the information $\sum_U \mathbf{x}_k^* = (N, N\bar{x}_U)'$. The final weights w_k obtained from (8.4) lead to the calibration estimator

$$\hat{Y}_{2B:WC/REG} = \frac{N}{\hat{N}} \sum_{p=1}^{P} F_p^{\circ} \left(\sum_{r_p} d_k (y_k - B_{2Br;d} x_k) \right) + N\bar{x}_U B_{2Br;d} \qquad (8.11)$$

with

$$B_{2Br;d} = \frac{\sum\limits_{p=1}^{P} F_p^{\circ} \left(\sum\limits_{r_p} d_k (x_k - \bar{x}_{r;w^{\circ}})(y_k - \bar{y}_{r;w^{\circ}}) \right)}{\sum\limits_{p=1}^{P} F_p^{\circ} \left(\sum\limits_{r_p} d_k (x_k - \bar{x}_{r;w^{\circ}})^2 \right)} \qquad (8.12)$$

where

$$\bar{y}_{r;w^{\circ}} = \frac{\sum\limits_{r} w_k^{\circ} y_k}{\sum\limits_{r} w_k^{\circ}} = \frac{\sum\limits_{p=1}^{P} F_p^{\circ} \left(\sum\limits_{r_p} d_k y_k \right)}{\hat{N}}$$

$\bar{x}_{r;w^{\circ}}$ is defined analogously, and $\hat{N} = \sum_{p=1}^{P} \hat{N}_p = \sum_s d_k$.

A quick examination of (8.7)–(8.12) shows that little separates the three estimator formulae. Apart from the presence in $\hat{Y}_{2B:WC/REG}$ of the factor $N/\hat{N}$, a minor difference lies in the weighting of the x and y data in the slope coefficients (8.8), (8.10) and (8.12). The factor $N/\hat{N}$ equals unity for SI and for STSI such that each group is also a stratum for sample selection.

On closer inspection, a difference of some significance can be seen. There is a contrast between $\hat{Y}_{SS:WC/REG}$ and $\hat{Y}_{2A:WC/REG}$ on the one hand and $\hat{Y}_{2B:WC/REG}$ on the other. The coefficients $B_{SSr;d}$ and $B_{2Ar;d}$ recognize separate group means, $\bar{y}_{r_p;d}$ and $\bar{x}_{r_p;d}$, whereas $B_{2Br;d}$ is expressed in terms of means $\bar{y}_{r;w^{\circ}}$ and $\bar{x}_{r;w^{\circ}}$ that are pooled over the whole response set. This has some impact on the prospects for reducing the nonresponse bias. All three use the available information in the calibrated weights computation, single-step is the simplest, computationally most direct procedure, and two-step B controls the final weights less rigorously than the other two.

A case of particular interest is STSI such that each stratum is also a group for calibration purposes. Each of the P groups indicated by $\boldsymbol{\gamma}_k = (\gamma_{1k}, \ldots, \gamma_{pk}, \ldots, \gamma_{Pk})'$ is also a stratum. The sampling weights in stratum p are $d_k = N_p/n_p$. We have $N/\hat{N} = 1$, and $F_p^\circ = n_p/m_p$, the inverse of the stratum response rate. The three estimators (8.7), (8.9) and (8.11), although still not identical, have the common form

$$\sum_{p=1}^{P} N_p(\bar{y}_{r_p} - B\bar{x}_{r_p}) + N\bar{x}_U B$$

where $\bar{y}_{r_p}$ and $\bar{x}_{r_p}$ are now straight respondent means in stratum p. The slope coefficient B has slightly different expressions in the three cases.

The estimators $\hat{Y}_{\text{SS:WC/REG}}$, $\hat{Y}_{\text{2A:WC/REG}}$ and $\hat{Y}_{\text{2B:WC/REG}}$ may differ slightly in their capacity to adjust for nonresponse bias. Depending on the data and on the response pattern, one of them may have a slight edge over the other two. In practice there are no rational grounds for deciding among them. A difference in the derivation is that $\hat{Y}_{\text{SS:WC/REG}}$ and $\hat{Y}_{\text{2A:WC/REG}}$ impose more extensive control on the weights than $\hat{Y}_{\text{2B:WC/REG}}$

All three stand in sharp contrast to calibration procedures that disregard some of the available information. They are likely to give estimators with significantly greater nonresponse bias. For the sake of comparison, let us consider the following example of an incomplete use of the available information.

Take $d_{\alpha k} = d_k$ as initial weights and use the auxiliary vector $\mathbf{x}_k = \mathbf{x}_k^* = (1, x_k)'$ with the information input $\sum_U \mathbf{x}_k = (N, X)'$. With final weights computed by (6.10), the resulting calibration estimator is $\hat{Y}_{\text{REG}}$ given by (7.12), which can also be written as

$$\hat{Y}_{\text{REG}} = \frac{N}{\hat{N}} \sum_r d_k(y_k - B_{r;d}x_k) + N\bar{x}_U B_{r;d} \tag{8.13}$$

with

$$B_{r;d} = \frac{\sum_r d_k(x_k - \bar{x}_{r;d})(y_k - \bar{y}_{r;d})}{\sum_r d_k(x_k - \bar{x}_{r;d})^2} \tag{8.14}$$

where $\bar{y}_{r;d} = \sum_r d_k y_k / \sum_r d_k$ and $\bar{x}_{r;d}$ is defined analogously. This procedure forgoes the information carried by the moon vector $\mathbf{x}_k^\circ = \boldsymbol{\gamma}_k$. As a result, formula (8.13) fails to recognize the grouping characteristic of the estimators for the fully-fledged single-step, two-step A and two-step B procedures. If pronounced differences exist between the group response propensities, $\hat{Y}_{\text{REG}}$ may incur a considerable bias, whereas (8.7), (8.9) and (8.11) have protection built in for such group differences.

Remark 8.1

The conditions of the example in this section were chosen to illustrate estimator formulae arising from a case of mild complexity in the auxiliary information. It is clear that for the considerably more complex $\mathbf{x}_k$-vectors that are used in practice, there is no prospect of obtaining simple closed expressions for the calibration estimator. They are not needed; the estimator is computed using available software. In the example, we have allowed the continuous variable x_k to be used in its original, continuous form. As argued elsewhere, a more prudent approach in practice is to use the known positive values $x_1, \ldots, x_N$ to form size groups, none of them too small, based on nonoverlapping intervals of the N values.   □

8.3. A MONTE CARLO SIMULATION STUDY OF ALTERNATIVE CALIBRATION PROCEDURES

The objective is to study the bias and the variance of the alternative calibration estimators in Section 8.2, derived by the single-step, two-step A and two-step B procedures. We obtain repeated samples and response sets from the KYBOK population of size $N = 832$ described in Section 7.7 and more particularly in Table 7.1. We use the same two response distributions as in Section 7.7, the logit and the increasing exponential, both with an average response rate of 86 %, and the same study variable y_k, expenditure on administration and maintenance.

The auxiliary information at the sample level is carried by the moon vector $\mathbf{x}_k^\circ = \boldsymbol{\gamma}_k$, where $\boldsymbol{\gamma}_k$ is the group identifier (8.5) with $P = 4$, indicating one of the four types of clerical municipality. The auxiliary information at the population level is carried by the star vector $\mathbf{x}_k^* = (1, x_k)'$ with the known total $(N, \sum_U x_k)$. As in Section 7.7, x_k is the square root of revenue advances. As in the preceding section we refer to the combined information as WC/REG.

The study includes the estimators $\hat{Y}_{\text{SS:WC/REG}}$, $\hat{Y}_{\text{2A:WC/REG}}$ and $\hat{Y}_{\text{2B:WC/REG}}$ given by (8.7), (8.9) and (8.11), respectively. The expansion estimator $\hat{Y}_{\text{EXP}} = (N/m) \sum_r y_k$, which uses no auxiliary information, is included as a benchmark with which the three more bias resistant alternatives can be compared. We know that $\hat{Y}_{\text{EXP}}$ will be heavily biased because it makes no attempt to adjust for the fact that the study variable in this example is significantly correlated with response propensity.

We drew 10 000 samples, each of size $n = 300$, by SI. For each sample, two response sets, one for each of the two response distributions, were obtained as described in Section 7.7. For each response set, the four estimators were computed, and if $\hat{Y}_{\text{W}}$ denotes one of the four estimators, the quantities $RB_{\text{SIM}}(\hat{Y}_{\text{W}})$ and $V_{\text{SIM}}(\hat{Y}_{\text{W}})$ were computed as in Section 7.7. The results are displayed in Table 8.1. The estimators are referred to by their abbreviated names SS:WC/REG, 2A:WC/REG, 2B:WC/REG and EXP.

The table confirms the anticipation that SS:WC/REG, 2A:WC/REG and 2B:WC/REG behave similarly with regard to both the absolute relative bias and the variance. The bias, negative for all three, is more pronounced for 2B:WC/REG than for the other two. A possible explanation is that the weights of 2B:WC/REG are less rigorously controlled: for this estimator, the calibration at the sample level, $\sum_r w_{2Bk} \boldsymbol{\gamma}_k = \sum_s d_k \boldsymbol{\gamma}_k$, does not hold. As expected, EXP shows much larger values of both relative bias and variance.

Table 8.1 Simulation relative bias (%) and simulation variance (multiplied by $\times\, 10^{-8}$) for two response distributions and four estimators. Samples of size $n = 300$ drawn by SI. Nonresponse rate 14 %.

Response distribution	Estimator	$RB_{\text{SIM}}(\hat{Y}_{\text{W}}) \times 10^2$	$V_{\text{SIM}}(\hat{Y}_{\text{W}}) \times 10^{-8}$
Logit	EXP	5.0	69.6
	SS:WC/REG	−0.6	9.7
	2A:WC/REG	−0.6	9.8
	2B:WC/REG	−0.8	9.5
Increasing exponential	EXP	9.3	70.1
	SS:WC/REG	−2.4	8.2
	2A:WC/REG	−2.3	8.3
	2B:WC/REG	−3.0	8.0

8.4.   TWO-STEP PROCEDURES IN PRACTICE

Of the single-step, two-step A and two-step B procedures, it is the last mentioned one that, more than the other two, conforms to a reasoning that is highly prevalent in practice. Let us consider this issue in more detail. In this reasoning, weighting must begin by adjusting the insufficiently large sampling weights for the nonresponse. Then, if applicable, the weights are made to conform to control totals known for the population. The first step is a nonresponse adjustment, and the second is a calibration on known population quantities whose primary goal is to achieve a consistency of the weight system. A secondary goal is to reduce the variance somewhat. Among important surveys steeped in this thinking are Statistics Canada's Labour Force Survey and the longitudinal surveys that depend thereon, and the Survey of Income and Program Participation (SIPP), an ongoing sample survey conducted by the US Bureau of the Census. Among the references that describe this methodology are Kalton and Kasprzyk (1986) and Rizzo et al. (1996).

There are two common outlooks on the initial step of adjusting the sampling weights:

- *Model fitting.* An explicit model is formulated, with the θ_k as unknown parameters. The model is fitted, $\hat{\theta}_k$ is obtained as an estimate of θ_k, and $1/\hat{\theta}_k$ is used as a weight adjustment to d_k.
- *Subgrouping.* The sample s is split into a number of subgroups. Several techniques are available for this. The inverse of the response fraction is then used as a weight adjustment to d_k for every member of the same group. The step is algorithmic; there is no explicit model formulation.

Logistic regression fitting is an example of the first category. The dependent variable in this fit is the dichotomous variable 'responding/not responding', with value $R_k = 1$ if $k \in r$ and $R_k = 0$ if $k \in s - r$. Fitting of this type was used, for example, by Folsom (1991), and by Ekholm and Laaksonen (1991) in examining nonresponse adjustment for the Finnish Household Budget Survey.

We can use maximum likelihood or some alternative method to fit the logistic regression model

$$\Pr(k \in r \,|\, s) = \theta_k = \frac{\exp(-(\mathbf{x}_k^\circ)'\boldsymbol{\beta})}{1 + \exp(-(\mathbf{x}_k^\circ)'\boldsymbol{\beta})} \tag{8.15}$$

using the data $(R_k, \mathbf{x}_k^\circ)$ for $k \in s$. This leads to an estimate $\hat{\boldsymbol{\beta}}$ of $\boldsymbol{\beta}$, and to estimated response probabilities for $k \in s$ given by

$$\hat{\theta}_k = \frac{\exp(-(\mathbf{x}_k^\circ)'\hat{\boldsymbol{\beta}})}{1 + \exp(-(\mathbf{x}_k^\circ)'\hat{\boldsymbol{\beta}})}. \tag{8.16}$$

The weights $w_k^\circ = d_k / \hat{\theta}_k$ with $\hat{\theta}_k$ given by (8.16) meet the objective of expanding the insufficiently large sampling weights d_k to an appropriate level, and there are grounds for believing that the estimator $\sum_r w_k^\circ y_k$ provides some protection against bias, especially if $\mathbf{x}_k^\circ$ is a powerful predictor of the response probability. All of these $\hat{\theta}_k$ belong to the unit interval, and all $w_k^\circ = d_k / \hat{\theta}_k$ are positive weights.

When star vector information is also present, $w_k^\circ = d_k / \hat{\theta}_k$ for $k \in r$ form a set of intermediate weights that can be employed as initial weights for the second step of two-step A or two-step B. In particular, when $\mathbf{x}_k^\circ = \boldsymbol{\gamma}_k$, the one-way classification vector, the maximum likelihood fit of (8.15) gives simply $\hat{\theta}_k = m_p / n_p$, where m_p is the number of respondents out of n_p sampled elements in group p. The resulting weight adjustment is $1/\hat{\theta}_k = n_p / m_p$ for all k in group p. By contrast, the first step of the two-step calibration, as discussed in Section 8.2, leads to the design-weighted adjustment $\sum_{s_p} d_k / \sum_{r_p} d_k$ for all k in group p.

There are several variations of the subgrouping method. It often consists in splitting the sample with the aid of a computerized algorithm into subgroups characterized by wide differences in response rate.

Although the method is often referred to as 'nonresponse modelling', there is usually no explicit model formulation, nor any reference to psychometric or other evidence substantiating the idea that the grouping finally settled on is a true explanation of how people react when in the process of completing a questionnaire.

The subgroup method assumes, more or less explicitly, that the nonresponse probability is the same for all elements within each group that happens to be the end result of the sample split. Such groups are often called *response homogeneity groups*. It must be kept in mind, however, that the grouping is sample-dependent and subject to randomness. For one given sample, two elements, say k and ℓ, may by chance end up in the same group. If a new sample happens to again include both k and ℓ, the computer algorithm may again by chance assign them to different groups, contradicting the idea that they have the same response probability.

One technique for forming the subgroups is to use the logistic regression estimates $\hat{\theta}_k$, obtained by (8.16) for a carefully chosen vector $\mathbf{x}_k$. The values $\hat{\theta}_k$ for $k \in s$ are placed into nonoverlapping subintervals of the unit interval, and the inverse of the response rate is used as a weight adjustment for all responding elements in one and the same group. This avoids very large weights $1/\hat{\theta}_k$

being obtained for elements with small fitted values $\hat{\theta}_k$. The method is discussed in Little (1986).

Stepwise methods for forming the groups include stepwise logistic regression fit and the segmentation method. These are used in longitudinal surveys at Statistics Canada, as reviewed in Dufour *et al.* (2001). References for the segmentation method are Kass (1980) and ANGOSS Software (1995). Both entail a stepwise selection of variables from a set of potential predictors of the response propensity. Each predictor variable is first transformed into a dichotomous analogue. For example, a continuous predictor variable value is treated as either 'high score' (with a dichotomized variable value set equal to 1) or 'low score' (value set equal to 0).

In the CHAID (Chi-square Automatic Interaction Detection) procedure, the sample is divided into subgroups using a set of dichotomous predictors. The grouping expands into a tree-like structure. An existing group is split into two subgroups that represent the widest possible separation in terms of response rate. A chi-square test is used to judge the significance of each new split. The segmentation continues until no further significant splits can be detected. A 'non-symmetric tree' is the result of the procedure. A review of subgrouping methods is found in Rizzo *et al.* (1996).

Splitting into too small groups can create instability. A threshold value should be defined, and no group size should be allowed to go below that value. We have emphasized that the calibration approach does not advocate an extremely fine split, into very small groups, of a given concept, such as age grouping or type of household. On the other hand, the calibration estimation technique does advocate the search for and incorporation of additional useful classifications into the $\mathbf{x}_k$-vector. One should thus add new important concepts for classification, rather than fragmenting an already existing classification. Support for this course of action will be found, for example, in Rizzo *et al.* (1996).

CHAPTER 9

Analysing the Bias due to Nonresponse

9.1. SIMPLE ESTIMATORS AND THEIR NONRESPONSE BIAS

In Chapter 7 we derived several simple special cases of the general calibration estimator $\hat{Y}_W$, whose weights w_k are given, under the standard specifications, by (6.10) and more generally by (6.14). The simulations in Sections 7.7 and 8.3 gave empirical evidence that the bias and variance are strongly related to the auxiliary information that is built into an estimator. The more extensive the auxiliary information, the lower the bias and the variance. The objective in this chapter is to examine the bias analytically. We shall examine the expression for the bias of $\hat{Y}_W$ and draw some conclusions.

The general definition for the bias of an estimator $\hat{Y}_W$ was given in Section 5.2. The bias of $\hat{Y}_W$, evaluated jointly with respect to the sampling design $p(s)$ and the response distribution $q(r|s)$, is

$$B_{pq}(\hat{Y}_W) = E_p\left(E_q(\hat{Y}_W|s)\right) - Y = E_{pq}(\hat{Y}_W) - Y.$$

The *relative bias* is given by

$$\text{relbias}(\hat{Y}_W) = B_{pq}(\hat{Y}_W)/Y.$$

In the simplest case, in Section 7.2, we have $\mathbf{x}_k = x_k = 1$ for all k, and $\sum_U x_k = N$. Then $\hat{Y}_W = \hat{Y}_{EXP} = N\bar{y}_{r;d}$, where $\bar{y}_{r;d} = \sum_r d_k y_k / \sum_r d_k$. A close approximation to the bias can easily be obtained. We replace the numerator sum and the denominator sum of $\bar{y}_{r;d}$ by their respective expected values. We have $E_{pq}\left(\sum_r d_k y_k\right) = E_p\left(\sum_s d_k \theta_k y_k\right) = \sum_U \theta_k y_k$ and $E_{pq}\left(\sum_r d_k\right) = \sum_U \theta_k$, so, to a close approximation,

$$B_{pq}(\hat{Y}_{EXP}) \approx N(\bar{y}_{U;\theta} - \bar{y}_U). \tag{9.1}$$

The approximate bias is thus proportional to the difference between two unknown means, the theta-weighted mean $\bar{y}_{U;\theta} = \sum_U \theta_k y_k / \sum_U \theta_k$ and the straight mean $\bar{y}_U = (1/N) \sum_U y_k$. The theta-weighting is indicated in $\bar{y}_{U;\theta}$ (and in similarly

Estimation in Surveys with Nonresponse   C.-E. Särndal and S. Lundström
© 2005 John Wiley & Sons, Ltd

weighted quantities) by the index θ, as explained in Remark 7.1. The relative bias of $\hat{Y}_{\text{EXP}}$ is

$$\text{relbias}(\hat{Y}_{\text{EXP}}) \approx \frac{\overline{y}_{U;\theta}}{\overline{y}_U} - 1. \tag{9.2}$$

The right-hand side can be positive or negative, depending on which of the two (unknown) means is greater, the theta-weighted one or the straight one. The relative bias is nearly zero if the response probabilities θ_k are constant throughout the population.

If we use also $(N-1)/N \approx 1$, (9.2) leads to another telling expression which links the relative bias to the coefficient of correlation between the response probability and the study variable,

$$\text{relbias}(\hat{Y}_{\text{EXP}}) \approx R_{\theta yU} cv_{\theta U} cv_{yU}$$

where the correlation coefficient is

$$R_{\theta yU} = \frac{\text{Cov}_{\theta yU}}{S_{\theta U} S_{yU}} \tag{9.3}$$

and the two coefficients of variation are

$$cv_{\theta U} = \frac{S_{\theta U}}{\overline{\theta}_U} \tag{9.4}$$

and

$$cv_{yU} = \frac{S_{yU}}{\overline{y}_U} \tag{9.5}$$

with $\overline{\theta}_U = \sum_U \theta_k / N$, $\text{Cov}_{\theta yU} = \sum_U (\theta_k - \overline{\theta}_U)(y_k - \overline{y}_U)/(N-1)$ and

$$S_{\theta U}^2 = \frac{\sum_U (\theta_k - \overline{\theta}_U)^2}{N-1}, \qquad S_{yU}^2 = \frac{\sum_U (y_k - \overline{y}_U)^2}{N-1}.$$

The greater the correlation $R_{\theta yU}$, the greater the relative bias, other things being equal. Even a modest correlation can result in a considerable relative bias. For example, if $R_{\theta yU} = 0.4$, $cv_{\theta U} = 0.5$ and $cv_{yU} = 1.5$, then $\text{relbias}(\hat{Y}_{\text{EXP}}) \approx 0.3$, which is unacceptably large.

For $\hat{Y}_{\text{EXP}}$ to be nearly unbiased, the requirements are exceptionally strong. They are met if the response probabilities θ_k are constant throughout the whole population, something highly unlikely. Most survey statisticians are well aware of this shortcoming of $\hat{Y}_{\text{EXP}}$. Although $\hat{Y}_{\text{EXP}}$ is not recommended for use, there is a certain interest in computing it in a survey, if only to see by how much it differs from alternative better estimates.

An effective type of auxiliary information is that which permits a grouping of the elements of the population or the sample. Ideally, the groups should be homogeneous with respect to the response probabilities and/or the study variable values. To establish such a grouping is not a trivial task. We examine this question in the next section.

The one-way classification discussed in Section 7.3 uses the group identifier $\boldsymbol{\gamma}_k = (\gamma_{1k}, \ldots, \gamma_{pk}, \ldots, \gamma_{Pk})'$ as an auxiliary vector. For InfoU with $\mathbf{x}_k =$

$\mathbf{x}_k^* = \boldsymbol{\gamma}_k$, we obtain the population weighting adjustment estimator $\hat{Y}_{\text{PWA}}$, given by (7.4). For InfoS with $\mathbf{x}_k = \mathbf{x}_k^{\circ} = \boldsymbol{\gamma}_k$, we obtain the weighting class estimator $\hat{Y}_{\text{WC}}$, given by (7.6). We approximate the bias by a reasoning similar to that used to obtain (9.1). We can alternatively obtain it as a special case of the much more general bias expression derived in Section 9.4. We have

$$B_{pq}(\hat{Y}_{\text{PWA}}) \approx B_{pq}(\hat{Y}_{\text{WC}}) \approx \sum_{p=1}^{P} N_p(\overline{y}_{U_p;\theta} - \overline{y}_{U_p}) \qquad (9.6)$$

where $\overline{y}_{U_p;\theta} = \sum_{U_p} \theta_k y_k / \sum_{U_p} \theta_k$ and $\overline{y}_{U_p} = (1/N_p) \sum_{U_p} y_k$. Thus $\hat{Y}_{\text{PWA}}$ and $\hat{Y}_{\text{WC}}$ share the same approximate bias. It is a function of the differences between theta-weighted and straight group means and is independent of the sampling design. For the relative bias we have

$$\text{relbias}(\hat{Y}_{\text{PWA}}) \approx \text{relbias}(\hat{Y}_{\text{WC}}) \approx \frac{\displaystyle\sum_{p=1}^{P} N_p \overline{y}_{U_p;\theta}}{\displaystyle\sum_{p=1}^{P} N_p \overline{y}_{U_p}} - 1. \qquad (9.7)$$

Alternatively, (9.6) can be written as

$$B_{pq}(\hat{Y}_{\text{PWA}}) \approx B_{pq}(\hat{Y}_{\text{WC}}) \approx \sum_{p=1}^{P} K_p \left(\sum_{U_p} \theta_k e_k \right) \qquad (9.8)$$

where $e_k = y_k - \overline{y}_{U_p}$ for $k \in U_p$, and $K_p = N_p / \sum_{U_p} \theta_k$. From (9.8) follow three conditions for nearly zero bias:

1 The response probabilities θ_k are constant within every group.
2 The residuals satisfy $e_k = y_k - \overline{y}_{U_p} = 0$ for all k within every group, that is, the y_k have zero variance within every group.
3 Neither 1 nor 2 holds necessarily, but $\sum_{U_p} \theta_k e_k = 0$ holds for every group, implying that θ_k and e_k are uncorrelated in every group.

Yet another way to express the bias is

$$B_{pq}(\hat{Y}_{\text{PWA}}) \approx B_{pq}(\hat{Y}_{\text{WC}}) \approx \sum_{p=1}^{P} N_p \overline{y}_{U_p} R_{\theta y U_p} c v_{\theta U_p} c v_{y U_p} \qquad (9.9)$$

where $R_{\theta y U_p}$, $cv_{\theta U_p}$ and $cv_{y U_p}$ are defined by analogy with (9.3)–(9.5), with U_p replacing U. Thus $R_{\theta y U_p}$ is the correlation within group p between the response probability and the y-variable. These group correlations play a key role in determining the bias. The bias can also be near zero if, fortuitously, the correlations $R_{\theta y U_p}$ have alternating signs for $p = 1, \ldots, P$, in such a way that positive and negative terms in (9.9) will roughly cancel each other.

9.2. FINDING AN EFFICIENT GROUPING

We continue the discussion of the important one-way classification, specified by the group identifier $\boldsymbol{\gamma}_k = (\gamma_{1k}, \ldots, \gamma_{pk}, \ldots, \gamma_{Pk})'$. A question arising is how to determine the P groups so as to reduce the nonresponse bias as far as possible.

The question has two aspects. For a given concept based on a continuous variable, such as age, we may ask how many age groups there should be, and how the age brackets should be defined. Too fine a division yields small group sizes, something that we may wish to avoid. If several categorical concepts are available – sex, education status, occupational group and so on – we can ask how many completely cross-classified concepts should enter into the vector $\boldsymbol{\gamma}_k$. For example, with two sexes, four classes of education status and five occupational groups, the cross-classification will give a vector $\boldsymbol{\gamma}_k$ of dimension $P = 40$. Again, we may wish to avoid too small group sizes.

The bias is nearly zero if either condition 1 or 2 of the preceding section holds. But in practice, one will most likely have to settle for groups within which some variability remains both in the y_k-values and in the response probabilities. We would like to identify groups that come close to fulfilling one or both of conditions 1 and 2. Let us consider some guidelines for this endeavour.

A grouping that fulfils condition 1 in Section 9.1 will result in a nearly zero bias for *all* of the perhaps numerous study variables in the survey. Attempts to meet condition 1 are particularly important. The condition says that the bias is nearly zero if the unknown response probabilities θ_k have zero variance within every group. We are led to examine the variability of the θ_k.

The total variation of the response probabilities can be decomposed as

$$\sum_U (\theta_k - \overline{\theta}_U)^2 = \sum_{p=1}^P \sum_{U_p} (\theta_k - \overline{\theta}_{U_p})^2 + \sum_{p=1}^P N_p (\overline{\theta}_{U_p} - \overline{\theta}_U)^2 \qquad (9.10)$$

where

$$\overline{\theta}_{U_p} = \frac{1}{N_p} \sum_{U_p} \theta_k \quad \text{and} \quad \overline{\theta}_U = \frac{1}{N} \sum_U \theta_k.$$

The left-hand side of (9.10) is an (unknown) population quantity independent of the grouping. A proposed grouping that brings about an increase in the between-groups component $\sum_{p=1}^P N_p (\overline{\theta}_{U_p} - \overline{\theta}_U)^2$ will cause an equivalent decrease in the within-groups component $\sum_{p=1}^P \sum_{U_p} (\theta_k - \overline{\theta}_{U_p})^2 = \sum_{p=1}^P (N_p - 1) S_{\theta U_p}^2$. The effect is that the within-group variance of θ_k, $S_{\theta U_p}^2 = \sum_{U_p} (\theta_k - \overline{\theta}_{U_p})^2 / (N_p - 1)$, will decrease for most or all groups. We come closer to fulfilling condition 1. If several groupings are compared, one can claim the best alternative to be the one that gives the largest between-groups component. Although (9.10) is stated in terms of unknown population quantities, we can estimate the between-groups component from the sample taken, as we now indicate.

Let $R_k, k \in s$, be the random variable defined by $R_k = 1$ if $k \in r$ and $R_k = 0$ if $k \in s - r$. The expectation of R_k, given s, is θ_k. Let $s_p = s \cap U_p$ be the sample obtained in group p. Then $\overline{R}_{s_p;d} = \sum_{s_p} d_k R_k / \sum_{s_p} d_k$ is nearly unbiased for $\overline{\theta}_{U_p}$, $p = 1, \ldots, P$, because

$$E_{pq}\left(\sum_{s_p} d_k R_k\right) = E_p\left(\sum_{s_p} d_k \theta_k\right) = \sum_{U_p} \theta_k, \qquad E_{pq}\left(\sum_{s_p} d_k\right) = N_p.$$

(The response probabilities are assumed here to depend on k but not on the sample s.) Similarly, $\overline{R}_{s;d} = \sum_s d_k R_k / \sum_s d_k$ is a nearly unbiased estimator of $\overline{\theta}_U$. Relying on the sample taken, we can, for any proposed grouping, compute

$$\sum_{p=1}^{P}\left(\sum_{s_p} d_k\right)(\overline{R}_{s_p;d} - \overline{R}_{s;d})^2 = \sum_{p=1}^{P}\frac{\left(\sum_{r_p} d_k\right)^2}{\sum_{s_p} d_k} - \frac{\left(\sum_r d_k\right)^2}{\sum_s d_k} \qquad (9.11)$$

where r_p is the set of responding elements in group p. The computation of (9.11) requires that the set s_p can be identified for $p = 1, \ldots, P$. That is, the value of the auxiliary vector $\mathbf{\gamma}_k = (\gamma_{1k}, \ldots, \gamma_{pk}, \ldots, \gamma_{Pk})'$ must be known for every element $k \in s$.

For example, in the case of SI with the sampling rate n/N, expression (9.11) becomes

$$\frac{N}{n}\left(\sum_{p=1}^{P}\frac{m_p^2}{n_p} - \frac{m^2}{n}\right) \qquad (9.12)$$

where n_p is the size of the part of the sample obtained in group p and m_p is the size of r_p.

The maximization of (9.11) or its special case (9.12) is a procedure that can be used in the search for an effective sample-based grouping s_p, $p = 1, \ldots, P$. The procedure reflects the ambition to use auxiliary variables to subdivide the sample into groups with the widest possible separation in terms of response propensity. This is the principle behind CHAID and other mechanical search algorithms (see Section 8.4).

The tool should be used with some caution. It would be misleading to claim that a grouping that maximizes (9.11) is 'optimal'. It must be remembered that the sample s is a random outcome. Over- or underrepresentation of groups occurs in s for purely random reasons. Therefore, information from other sources should be used whenever available, for example, evidence from other surveys about population groups that may have atypical response rates.

Condition 2 in Section 9.1 says that a nearly zero bias is achieved if $e_k = y_k - \overline{y}_{U_p} = 0$ for all k within every group, that is, if the y_k-values are constant within groups. Most surveys involve several (or even many) study variables. Thus, it is difficult or impossible to find groups such that the y_k-values are constant not only for every one of the groups, but also for every one of the y-variables. One would have to rely on professional judgement and experience to identify the most important study variables and then attempt to meet condition 2 for those variables.

Section 9.1 showed the bias for a few simple cases of the calibration estimator. We need a more general examination. This is accomplished in Sections 9.4 and 9.5. We find that the better one can succeed in incorporating relevant auxiliary information into the $\mathbf{x}_k$-vector, the better, generally speaking, are the chances of limiting the nonresponse bias.

Building a potent $\mathbf{x}_k$-vector becomes of paramount importance. The statistician must first make a thorough inventory of all potential auxiliary variables. This process, which may involve a matching of administrative registers, may reveal a perhaps surprisingly large number of potential auxiliary variables. This step should be followed by a selection of the most pertinent variables. Principles to guide this effort have already been mentioned in Section 3.3, and we shall dwell on them in more detail in Chapter 10.

9.3. FURTHER ILLUSTRATIONS OF THE NONRESPONSE BIAS

For another illustration of the bias due to nonresponse, we return to the comparison in Section 8.2 between the single-step, two-step A and two-step B calibration procedures. The information referred to as WC/REG was defined in Section 8.2. The information at the sample level permits every $k \in s$ to be placed in the appropriate group indicated by the moon vector $\mathbf{x}_k^\circ = \boldsymbol{\gamma}_k = (\gamma_{1k}, \dots, \gamma_{pk}, \dots, \gamma_{Pk})'$. The information at the population level is transmitted by the star vector $\mathbf{x}_k^* = (1, x_k)'$ with the known total $\sum_U \mathbf{x}_k^* = (N, \sum_U x_k)'$.

The estimators $\hat{Y}_{\text{SS:WC/REG}}$, $\hat{Y}_{\text{2A:WC/REG}}$ and $\hat{Y}_{\text{2B:WC/REG}}$ were given by (8.7), (8.9) and (8.11), respectively. How do they compare in their capacity to control the bias caused by nonresponse? The empirical evidence from the Monte Carlo study in Section 8.3 suggests that there are small differences only with regard to bias. Further insight is obtained in this section by examining the analytic expressions for the bias.

We obtain the approximate bias for the three estimators by a special analysis of each estimator. The details are omitted. In particular, the bias expression for $\hat{Y}_{\text{SS:WC/REG}}$ can be obtained by use of Corollary 9.1 with $\mathbf{z}_k = \mathbf{x}_k = (x_k, \boldsymbol{\gamma}_k')'$. If $\hat{Y}$ stands for one of the three estimators, the common expression for their approximate bias is

$$B_{pq}(\hat{Y}) \approx \sum_{p=1}^{P} N_p[(\overline{y}_{U_p;\theta} - \overline{y}_{U_p}) - B(\overline{x}_{U_p;\theta} - \overline{x}_{U_p})] \qquad (9.13)$$

where N_p is the size of group p, $\overline{y}_{U_p;\theta} = \sum_{U_p} \theta_k y_k / \sum_{U_p} \theta_k$ and $\overline{y}_{U_p} = \sum_{U_p} y_k / N_p$. The x-means $\overline{x}_{U_p;\theta}$ and $\overline{x}_{U_p}$ have analogous expressions. Only the coefficient B differs from one estimator to the next. The quantities in (9.13) are population quantities, usually unknown, and an attempt to estimate (9.13) runs up against the predicament of unknown θ_k.

For the single-step estimator $\hat{Y}_{SS:WC/REG}$, we have

$$B = B_{SS} = \frac{\sum\limits_{p=1}^{P} \sum\limits_{U_p} \theta_k (x_k - \bar{x}_{U_p;\theta})(y_k - \bar{y}_{U_p;\theta})}{\sum\limits_{p=1}^{P} \sum\limits_{U_p} \theta_k (x_k - \bar{x}_{U_p;\theta})^2};$$

for the two-step A estimator $\hat{Y}_{2A:WC/REG}$,

$$B = B_{2A} = \frac{\sum\limits_{p=1}^{P} K_p \sum\limits_{U_p} \theta_k (x_k - \bar{x}_{U_p;\theta})(y_k - \bar{y}_{U_p;\theta})}{\sum\limits_{p=1}^{P} K_p \sum\limits_{U_p} \theta_k (x_k - \bar{x}_{U_p;\theta})^2};$$

and for the two-step B estimator $\hat{Y}_{2B:WC/REG}$,

$$B = B_{2B} = \frac{\sum\limits_{p=1}^{P} K_p \sum\limits_{U_p} \theta_k (x_k - \bar{\bar{x}}_{U;\theta})(y_k - \bar{\bar{y}}_{U;\theta})}{\sum\limits_{p=1}^{P} K_p \sum\limits_{U_p} \theta_k (x_k - \bar{\bar{x}}_{U;\theta})^2}$$

where $\bar{\bar{y}}_{U;\theta} = (1/N) \sum_{p=1}^{P} N_p \bar{y}_{U_p;\theta}$, the expression for $\bar{\bar{x}}_{U;\theta}$ is analogous, and $K_p = N_p / \sum_{U_p} \theta_k$. We can also write (9.13) as

$$B_{pq}(\hat{Y}) \approx \sum_{p=1}^{P} K_p \left(\sum_{U_p} \theta_k e_k \right)$$

where $e_k = y_k - \bar{y}_{U_p} - B(x_k - \bar{x}_{U_p})$. Compared to (9.8), only the residuals e_k have changed. For each of the estimators, we note three sufficient conditions for a nearly zero bias:

1 The response probabilities θ_k are constant within every group, because if they are, then $\bar{y}_{U_p;\theta} - \bar{y}_{U_p} = \bar{x}_{U_p;\theta} - \bar{x}_{U_p} = 0$ for every group $p = 1, \ldots, P$.
2 The residuals are such that $e_k = y_k - \bar{y}_{U_p} - B(x_k - \bar{x}_{U_p}) = 0$ for all k within every group.
3 $\sum_{U_p} \theta_k e_k = 0$ holds for every group, as is the case when θ_k and e_k are uncorrelated in every group.

Suppose that the relation between the y-variable and the x-variable is such that $y_k = \alpha_p + \beta x_k$ holds for $k \in U_p$, where the intercepts α_p vary between groups while β is a common slope. Then condition 2 is satisfied, and the bias is nearly zero for $\hat{Y}_{SS:WC/REG}$ and $\hat{Y}_{2A:WC/REG}$, but not necessarily for $\hat{Y}_{2B:WC/REG}$. To conclude that the bias is nearly zero for $\hat{Y}_{2B:WC/REG}$, we need the stronger condition $y_k = \alpha + \beta x_k$ for $k \in U$, with an intercept α that is common to all groups. This suggests that $\hat{Y}_{2B:WC/REG}$ may be more exposed to bias than $\hat{Y}_{SS:WC/REG}$ and $\hat{Y}_{2A:WC/REG}$. However, the difference in bias is expected to be minor for many practical purposes.

9.4. A GENERAL EXPRESSION FOR THE BIAS OF THE CALIBRATION ESTIMATOR

Sections 9.1 and 9.3 exemplified the bias caused by nonresponse under specific formulations of the auxiliary vector. It became clear that the conditions for a nearly zero bias are related to certain properties of the θ_k and to certain properties of the residuals of the regression of y_k on x_k. We need a more thorough examination of the basic properties of the calibration estimator $\hat{Y}_W$.

Both the bias caused by nonresponse and the variance are important aspects of $\hat{Y}_W$. Of these, the bias must be our main concern. Our first objective is to limit the bias as far as possible. To minimize variance is of secondary importance, because if an estimator is greatly biased, it is poor consolation that its variance is low. The mean squared error may be large. For estimates at the level of the whole population, the variance is often small, but the bias can be considerable. In estimation for domains, the variance can also be large. We discuss the nonresponse bias in this and the next chapter. The variance is examined in Chapter 11.

We examine the (single-step) calibration estimator $\hat{Y}_W = \sum_r w_k y_k$, where the weights w_k are given by (6.14) with $d_{\alpha k} = d_k = 1/\pi_k$ as initial weights. For InfoUS, the auxiliary vector is $\mathbf{x}_k = \begin{pmatrix} \mathbf{x}_k^* \\ \mathbf{x}_k^\circ \end{pmatrix}$ and the corresponding information input is $\mathbf{X} = \begin{pmatrix} \sum_U \mathbf{x}_k^* \\ \sum_s d_k \mathbf{x}_k^\circ \end{pmatrix}$. The detailed features of the information are specified in Section 6.2. The instrument vector is $\mathbf{z}_k$. It is allowed to differ from $\mathbf{x}_k$, but the standard specification is $\mathbf{z}_k = \mathbf{x}_k$. The weights are given by (8.1) when $\mathbf{z}_k = \mathbf{x}_k$. The special cases InfoU and InfoS are covered by letting $\mathbf{x}_k = \mathbf{x}_k^*$ and $\mathbf{x}_k = \mathbf{x}_k^\circ$, respectively.

It is not possible to obtain an exact and explicit expression showing how the bias depends on the chosen auxiliary vector. But a close approximation can be obtained. We call this close approximation the *nearbias* of $\hat{Y}_W$. It is given in the following proposition.

Proposition 9.1

For large response sets, the bias of the calibration estimator $\hat{Y}_W$ defined by (6.14) with $d_{\alpha k} = d_k = 1/\pi_k$ is given approximately by

$$B_{pq}(\hat{Y}_W) = E_{pq}(\hat{Y}_W) - Y \approx \text{nearbias}(\hat{Y}_W)$$

where

$$\text{nearbias}(\hat{Y}_W) = -\sum_U (1 - \theta_k) e_{\theta k} \tag{9.14}$$

with

$$e_{\theta k} = y_k - \mathbf{x}_k' \mathbf{B}_{U;\theta}, \qquad \mathbf{B}_{U;\theta} = \left(\sum_U \theta_k \mathbf{z}_k \mathbf{x}_k' \right)^{-1} \sum_U \theta_k \mathbf{z}_k y_k. \qquad (9.15)$$

□

Expression (9.14) is given in Lundström (1997). Other authors have derived similar expressions, under different conditions and assumptions. Bethlehem (1988) examines regression-based estimators for nonresponse; they bear resemblance to the calibration estimators considered here. He arrives at a related expression for the bias.

The end-of-chapter appendix presents the derivation of (9.14). The steps of the derivation are not essential for an understanding of the following sections. The technique is to isolate the principal term of the bias and to drop the rest, a lower-order term tending to zero when the size of the response set r increases. The nearbias $- \sum_U (1 - \theta_k) e_{\theta k}$ will closely approximate $B_{pq}(\hat{Y}_W)$ even for response sets of rather modest size. In the following we shall analyse the nearbias, in general and for specific $\mathbf{x}_k$-vectors. It is zero under certain conditions. When they hold, it is fitting to call $\hat{Y}_W$ *nearly unbiased*.

Remark 9.1

The condition on the instrument vector $\mathbf{z}_k$ mentioned in Remark 6.8 plays an important role in this and following chapters. The condition is that $\boldsymbol{\mu}' \mathbf{z}_k = 1$ holds for all $k \in U$, for some constant, nonrandom vector $\boldsymbol{\mu}$. For example, if $\mathbf{z}_k = \mathbf{x}_k = \boldsymbol{\gamma}_k$, where $\boldsymbol{\gamma}_k = (\gamma_{1k}, \ldots, \gamma_{Pk})'$ is the one-way classification vector, then $\boldsymbol{\mu}' \mathbf{z}_k = 1$ holds for all $k \in U$ by taking $\boldsymbol{\mu}$ to be the P-vector whose entries are all equal to one.   □

Remark 9.2

Several interesting conclusions follow from Proposition 9.1:

- We expect $\hat{Y}_W$ to be nearly unbiased if the response probability θ_k is constant throughout, that is, if $\theta_k = \theta$ for all $k \in U$. This expectation is confirmed if $\boldsymbol{\mu}' \mathbf{z}_k = 1$ holds for all k, because then $B_{pq}(\hat{Y}_W) \approx -(1 - \theta) \sum_U e_{\theta k} = 0$. In particular, if $\mathbf{z}_k = \mathbf{x}_k$ and $\boldsymbol{\mu}' \mathbf{x}_k = 1$ for all k, then the nearbias is zero if θ_k is constant throughout.
- The bias of $\hat{Y}_W$, or any alternative estimator, is not zero when there is non-response, except in unlikely circumstances. The bias of $\hat{Y}_W$ is caused by the nonresponse, not by the probability sampling. We are led to ask what survey conditions and what relations among variables must hold for the bias to be nearly zero. To the best of our judgement, do we come close to realizing those conditions in our survey? We dwell on these questions in Chapter 10.
- The nearbias $- \sum_U (1 - \theta_k) e_{\theta k}$ is independent of the sampling design used to draw s, even though the point estimator $\hat{Y}_W = \sum_r w_k y_k$ depends on the design through the sampling weights $d_k = 1/\pi_k$. Whatever the sampling design,

the nearbias is the same as long as the $\mathbf{x}_k$-vector is the same. By contrast, the nearbias depends on the response distribution and its unknown response probabilities.

- Let $\mathbf{x}_{0k}$ be a vector with a fixed composition of auxiliary variables. Then we conclude from (9.14) that the nearbias is the same for $\mathbf{x}_k = \mathbf{z}_k = \mathbf{x}_k^\circ = \mathbf{x}_{0k}$ (InfoS; information on $\mathbf{x}_{0k}$ up to s only) as it is for $\mathbf{x}_k = \mathbf{z}_k = \mathbf{x}_k^* = \mathbf{x}_{0k}$ (InfoU; information on $\mathbf{x}_{0k}$ up to the population U). In other words, the nearbias would not have been reduced any further had the information about an x-variable in $\mathbf{x}_k^\circ = \mathbf{x}_{0k}$ extended all the way up to the level U (in which case that variable would qualify for inclusion in $\mathbf{x}_k^*$). An auxiliary variable is just as effective as a bias-reducing tool whether it carries information up to U or 'only' up to s. We have already noted an example of this in discussing the one-way classification in Section 9.1. The nearbias of $\hat{Y}_{\mathrm{PWA}}$ is the same as that of $\hat{Y}_{\mathrm{WC}}$ and is given by (9.8). (The nearbias is a first-order, but close, approximation. The *exact* bias differs slightly between InfoS and InfoU, as explained in the end-of-chapter appendix.)

  □

Proposition 9.1 emphasizes the need to identify powerful auxiliary variables carrying information *at least* up to the level of s. For some variables, it may be impractical or too costly in a survey to seek the more extensive information extending up to the level of the whole population. But this effort is not necessary if the principal objective is to control the bias. To have information up to s is sufficient. The perspective is somewhat different if we judge an auxiliary variable for its potential to reduce variance. Then it is a definite advantage if the information extends up to U. But bias, not variance, is our principal concern.

The following corollary follows easily from Proposition 9.1 by noting that if $\boldsymbol{\mu}'\mathbf{z}_k = 1$ for all k, then $\sum_U \theta_k e_{\theta k} = \boldsymbol{\mu}' \sum_U \mathbf{z}_k \theta_k e_{\theta k} = 0$.

Corollary 9.1

If $\boldsymbol{\mu}'\mathbf{z}_k = 1$ for all k, then the nearbias of $\hat{Y}_{\mathrm{W}}$ in (9.14) has three equivalent expressions:

$$\mathrm{nearbias}(\hat{Y}_{\mathrm{W}}) = -\sum\nolimits_U e_{\theta k} \tag{9.16}$$

where $e_{\theta k}$ is given by (9.15);

$$\mathrm{nearbias}(\hat{Y}_{\mathrm{W}}) = \left(\sum\nolimits_U \mathbf{x}_k \right)' (\mathbf{B}_{U;\theta} - \mathbf{B}_U) \tag{9.17}$$

where $\mathbf{B}_U = \left(\sum_U \mathbf{z}_k \mathbf{x}_k' \right)^{-1} \left(\sum_U \mathbf{z}_k y_k \right)$; and

$$\mathrm{nearbias}(\hat{Y}_{\mathrm{W}}) = \sum\nolimits_U \mathbf{x}_k' \mathbf{B}_{e:U;\theta} \tag{9.18}$$

where $\mathbf{B}_{e:U;\theta} = \left(\sum_U \theta_k \mathbf{z}_k \mathbf{x}_k' \right)^{-1} \left(\sum_U \theta_k \mathbf{z}_k e_k \right)$ and $e_k = y_k - \mathbf{x}_k' \mathbf{B}_U$. These results hold independently of the sampling design.   □

An interesting feature of (9.17) is that it shows the nearbias of $\hat{Y}_{\mathrm{W}}$ as a function of the difference between two regression vectors, the theta-weighted $\mathbf{B}_{U;\theta}$

and the equal-weighted $\mathbf{B}_U$. Although both are regression vectors, they can differ considerably, causing considerable bias. A simple illustration of (9.17) is the result (9.1) for InfoU with $\mathbf{x}_k = \mathbf{z}_k = 1$ for all k. In this case, $\sum_U \mathbf{x}_k = N$, $\mathbf{B}_{U;\theta} = \overline{y}_{U;\theta}$ and $\mathbf{B}_U = \overline{y}_U$. Another simple illustration of (9.17) is (9.6), obtained under the specifications $\mathbf{x}_k = \mathbf{z}_k = \mathbf{x}_k^* = \boldsymbol{\gamma}_k$ (for InfoU) and $\mathbf{x}_k = \mathbf{z}_k = \mathbf{x}_k^\circ = \boldsymbol{\gamma}_k$ (for InfoS).

9.5. CONDITIONS FOR NEAR-UNBIASEDNESS

We pursue the analysis of the nonresponse bias of the single-step calibration estimator $\hat{Y}_W$ for InfoUS. There is information at the sample level combined with information at the population level. The weight system is given by (6.14) with $d_{\alpha k} = d_k$; the instrument vector $\mathbf{z}_k$ is often but not necessarily equal to $\mathbf{x}_k$. The special cases InfoU and InfoS are also covered. We shall see that it depends on key properties of the auxiliary vector $\mathbf{x}_k$ whether or not the bias of $\hat{Y}_W$ can be essentially eliminated.

Expression (9.14) for the nearbias of $\hat{Y}_W$ is valid for any sampling design. When $\boldsymbol{\mu}'\mathbf{z}_k = 1$ for all k, we know from Remark 9.2 that the nearbias is zero if the response probability θ_k is constant throughout the population. This is a severe requirement. But the nearbias can be zero under less demanding conditions on the response distribution.

Proposition 9.2 shows that the nearbias is zero if a linear relation exists between the inverse of the response probability $1/\theta_k$ and the instrument $\mathbf{z}_k$. Therefore the quantities $1/\theta_k$ will play an important role in the following. It is convenient to introduce a special name and notation. We use $\phi_k = 1/\theta_k$ and call ϕ_k the *response influence*, or simply the *influence* of element k. An element has a high response influence if it has a low response probability, just as it has a high design weight $d_k = 1/\pi_k$ if it has a low probability of being sampled. An ordinary dictionary considers 'weight' and 'influence' to be essentially synonymous words, but there is some difference. In our context, the difference is that a weight, such as d_k, is known and computable, whereas the influence ϕ_k is unknown. We shall use a technique for replacing ϕ_k by computable entities, fit for use in estimation.

In particular, when $\mathbf{z}_k = \mathbf{x}_k$, the nearbias is zero when ϕ_k and $\mathbf{x}_k$ are linearly related. The implication for practice is that among possible alternatives for $\mathbf{x}_k$, we should seek one for which this linear relationship holds.

Proposition 9.2

The nearbias (9.14) of $\hat{Y}_W$ is zero if the response influence has the form

$$\phi_k = 1 + \boldsymbol{\lambda}'\mathbf{z}_k, \qquad \text{for } k \in U \tag{9.19}$$

and for some constant, nonrandom vector $\boldsymbol{\lambda}$ not dependent on k. Furthermore, if $\boldsymbol{\mu}'\mathbf{z}_k = 1$ for all k, then (9.16)–(9.18) in Corollary 9.1 apply and the nearbias is zero whenever the response influence has the form

$$\phi_k = \boldsymbol{\lambda}'\mathbf{z}_k, \qquad \text{for } k \in U \tag{9.20}$$

and for some constant, nonrandom vector λ not dependent on k. Under the standard specification $\mathbf{z}_k = \mathbf{x}_k$, condition (9.19) reads

$$\phi_k = 1 + \lambda'\mathbf{x}_k, \qquad \text{for } k \in U \qquad (9.21)$$

and (9.20) reads

$$\phi_k = \lambda'\mathbf{x}_k, \qquad \text{for } k \in U. \qquad (9.22)$$

These results hold for any sampling design. □

Since the influence ϕ_k belongs to the interval $[1, \infty)$ for all k, the vector λ should be one that respects this requirement. That λ is constant means that it has the same value for all k; that it is nonrandom means that it must not depend on random sets such as s and r.

Proof. When (9.19) holds, then $1 - \theta_k = \theta_k\lambda'\mathbf{z}_k$ holds for $k \in U$, and therefore, from (9.14),

$$\text{nearbias}(\hat{Y}_W) = -\sum_U (1 - \theta_k)e_{\theta k} = -\lambda'\sum_U \theta_k\mathbf{z}_k(y_k - \mathbf{x}_k'\mathbf{B}_{U;\theta}) = 0.$$

The proofs of the other parts of the proposition are analogous. □

Condition (9.19) is given in Lundström (1997). Fuller *et al.* (1994) consider estimators related to the calibration estimators examined here. They obtain a condition for nearly zero bias resembling (9.20).

Proposition 9.2 is illustrated by several examples in Chapter 10. When $\mathbf{z}_k = \mathbf{x}_k$, the proposition implies that we must seek an auxiliary vector $\mathbf{x}_k$ that improves as far as possible the chances of fulfilling condition (9.21) or (9.22).

The instrument vector $\mathbf{z}_k$ is at our disposal; we can specify it to suit our objectives. Thus even after fixing the auxiliary vector $\mathbf{x}_k$, there exists, as a last resource, one possibility for controlling the bias, namely, by an appropriate choice of $\mathbf{z}_k$. We illustrate this possibility in Section 10.5.

The composition of the auxiliary vector is the key to a successful estimation in the presence of nonresponse. Proposition 9.2 deals with desirable links between the response influence and the auxiliary vector (or the instrument vector, if different). The following proposition deals with desirable links between the study variable and the auxiliary vector.

Proposition 9.3

The nearbias (9.14) of $\hat{Y}_W$ is zero if the study variable values have the form

$$y_k = \boldsymbol{\beta}'\mathbf{x}_k, \qquad \text{for } k \in U \qquad (9.23)$$

and for some constant, nonrandom vector $\boldsymbol{\beta}$ not dependent on k. □

Proof. When (9.23) holds for all k, then $e_{\theta k}$ in (9.15) is zero for all k, and therefore the nearbias (9.14) is zero. □

The perfect relationship between y_k and $\mathbf{x}_k$ expressed by (9.23) will not hold in practice. However, we can expect the bias of $\hat{Y}_W$ to be small if that relationship comes close to being attained. We should strive for a vector $\mathbf{x}_k$ such that the residuals $e_k = y_k - \mathbf{x}_k' \mathbf{B}_U$ with $\mathbf{B}_U = \left(\sum_U \mathbf{z}_k \mathbf{x}_k' \right)^{-1} \left(\sum_U \mathbf{z}_k y_k \right)$ are small. Then both the bias and the variance of the calibration estimator will be small.

Propositions 9.2 and 9.3 form the point of departure for the procedures in Chapter 10, which aim to identify a suitable $\mathbf{x}_k$-vector for the calibration estimator.

9.6. A REVIEW OF CONCEPTS, TERMS AND IDEAS

In this chapter we have seen several illustrations of the fact that the nonresponse bias is linked to relations that may or may not exist between three types of entities: the response probability, the study variable and the auxiliary vector. Statisticians have long been interested in how, more precisely, the bias depends on these perceived relations. As a result, important work in the past few decades has sought to give structure to nonresponse theory with the aid of such relations.

A plethora of terms and concepts has evolved in this regard. These terms and concepts are frequently seen in the literature and are often heard in discussions among statisticians. They have precise mathematical definitions, at least in their original formulations. In rapid conversation, however, their precise meanings are sometimes glossed over.

Because the terms are frequently mentioned and used, let us see how they affect the material in this book. Terms relating to the failure to observe some of the desired y-data are missing at random (MAR) and missing completely at random (MCAR). Terms referring to the nature of a perceived response mechanism are ignorable, nonignorable and unconfounded.

These terms are found in important sources such as Rubin (1983, 1987) and Little and Rubin (1987). Rubin (1992) points out that his thinking was, at an early stage at least, motivated by Bayesian notions. For example, 'ignorable mechanism' really meant 'Bayesianly ignorable mechanism'. The terms have been interpreted by other authors to fit other modes of inference, not necessarily Bayesian.

In the context of frequentistic survey sampling, the terms are also used in ways that may differ somewhat from the original formulations. The terms can have some value in communicating ideas and distinctions with regard to the nonresponse problem.

For the terms MCAR, MAR (which are cases of ignorable missingness) and nonignorable missingness, we cite descriptions given in Bethlehem (1999) and Lohr (1999). They are fairly representative of statements in the literature of the last 15 years. As usual, y denotes the study variable, x an auxiliary variable, and $\mathbf{x}$ an auxiliary vector.

Bethlehem (1999) describes MCAR and MAR as follows. MCAR occurs when 'the probability of a missing value for y is independent of the value of y and independent of the value of x. . . . Then the observed values of y form a random sub-sample from the sample.' MAR occurs when the probability of a missing value for y is independent of the value of y but may depend on the value of x. 'Then the observed values of y do not form a sub-sample of the sample. However, they are a random sub-sample within the classes defined by the values of x. The auxiliary variable can be used to effectively correct for a bias due to missing values.'

Lohr's (1999) similar frequentistic interpretation maintains (in our notation) that if the response probability θ_k depends on $\mathbf{x}_k$ but not on y_k, the data are MAR. 'The nonresponse depends only on observed variables. We can successfully model the nonresponse, since we know the values of $\mathbf{x}_k$ for all sample elements.'

On closer examination, these verbal descriptions of MAR and MCAR are not so simple. Different interpretations are perhaps possible, but the most direct ones are not without certain ambiguities. There is a finite population U, and associated with an element $k \in U$ is the triple $(\theta_k, y_k, \mathbf{x}_k)$. It is rarely realistic to imagine that θ_k depends on $\mathbf{x}_k$ but not on y_k. Suppose that θ_k and $\mathbf{x}_k$ are strongly related, while at the same time θ_k and y_k are unrelated. The chances are then that $\mathbf{x}_k$ and y_k are virtually unrelated. But that would go contrary to our preference for a strong relation between $\mathbf{x}_k$ and y_k, a relation which is essential both in order to control the nonresponse bias and to reduce the variance of an estimator of the y-total $Y = \sum_U y_k$.

On the other hand, if θ_k and $\mathbf{x}_k$ are indeed strongly related, a not uncommon situation is that $\mathbf{x}_k$ and y_k are also related, perhaps strongly so. To imagine then that θ_k and y_k are unrelated is counterintuitive and unlikely to hold in a finite population.

One can hardly ignore the fact that, almost without exception, a relation exists between θ_k and at least some of the study variables, of which the survey may have many. The estimators, for all y-variables, must strive to take this relation into account, in the best possible manner, and this is done by the use of auxiliary information. Otherwise, the survey results may be seriously biased.

As defined by Lohr (1999), *ignorable nonresponse* means that 'a model can explain the nonresponse mechanism and the nonresponse can be ignored after the model accounts for it, not that the nonresponse can be completely ignored and complete data methods used'. The decisive point here is whether or not a chosen model is capable of explaining the nonresponse mechanism.

How do the above statements relate to the reasoning in this book, which advocates calibration as the principal technique? Calibration produces weights based on a given input of information and without any model statement. Suppose that instead of calibration, we develop an estimator from a set of assumptions and model statements. Suppose the survey conditions allow a one-way classification of the sample elements into groups indexed $p = 1, \ldots, P$. Perhaps hard pressed for better alternatives, the statistician may venture the assumption that the response mechanism is explained by the model postulating that the elements in r_p (the respondents in class p) are a random subsample of the elements in s_p (the sampled elements in group p). Then the probability of response, given s_p, is the same for all elements in s_p. We let this model 'account for the nonresponse', then invoke the idea that 'nonresponse can be ignored'. We are thus content to compute survey estimates using equal weights for all responding elements in group

p, namely, Nn_p/nm_p under SI, or more generally $d_k n_p/m_p$, where $d_k = 1/\pi_k$. Both weightings conform to the idea of 'a random subsampling within classes', with m_p respondents drawn randomly from n_p in the group s_p. The estimator of $Y = \sum_U y_k$ becomes $\hat{Y} = \sum_{p=1}^P n_p \left(\sum_{r_p} d_k y_k \right) /m_p$. Further modelling is perhaps not attempted. But if pressed for an opinion, no statistician truly believes that the responding elements in group p are a random subsample or that they have equal response probability, which is the condition under which $\hat{Y}$ becomes nearly free of bias.

In the calibration approach this case is resolved by specifying the auxiliary vector as $\mathbf{x}_k = \mathbf{x}_k^\circ = \boldsymbol{\gamma}_k$, as in Section 7.3. The calibrated weights are $w_k = d_k \sum_{s_p} d_k / \sum_{r_p} d_k$. They give the calibration estimator $\hat{Y}_{\text{WC}} = \sum_{p=1}^P \hat{N}_p \bar{y}_{r_p;d}$ with $\hat{N}_p = \sum_{s_p} d_k$ and $\bar{y}_{r_p;d} = \sum_{r_p} d_k y_k / \sum_{r_p} d_k$. It respects the sampling design, with d-weighting in all relevant places. The weights are produced without any assumption that the respondents are a random subsample within every group. But when subjected to analysis, $\hat{Y}_{\text{WC}}$ is found to be nearly without bias if $\theta_k = a_p$, a constant, for all k in group p, for $p = 1, \ldots, P$.

The weights $d_k \sum_{s_p} d_k / \sum_{r_p} d_k$ (from the calibration approach) and $d_k n_p/m_p$ (from the modelling argument) do not agree, except in special cases, as when STSI is used and each stratum coincides with one of the groups indicated by $\boldsymbol{\gamma}_k$.

Commenting on the concept nonignorable nonresponse, Bethlehem (1999) states: 'The probability of a missing value for y depends both on the value of y and on the value of x. Then the observed values of y do not form a random sub-sample of the sample. Also, they do not form a random sub-sample within the classes defined by the values of x. Therefore the auxiliary variable cannot be used effectively to correct for a bias due to missing values.' In a similar vein, Lohr (1999) defines nonignorable nonresponse as follows: 'If the probability of nonresponse depends on the value of a response variable and cannot be completely explained by values of the $\mathbf{x}$'s, then the nonresponse is non-ignorable.'

By comparison, the position in this book is that θ_k and y_k are likely to always be related to some extent, and that 'complete explanation' of θ_k in terms of a vector $\mathbf{x}_k$ is virtually never possible. All situations are nonignorable. We can never hope to have an auxiliary vector liable to correct, with full effectiveness, for bias due to nonresponse. Expressed somewhat differently, if we insist on using the word 'model': no model, regardless of what $\mathbf{x}_k$-vector it is built on, can completely explain the response probabilities θ_k. A relationship between θ_k and $\mathbf{x}_k$ is more or less clearly pronounced, and we can measure it from a sample, but to hope that this $\mathbf{x}_k$ will achieve a 'complete explanation' of θ_k is utopian.

Important as the terms MAR, MCAR, ignorable and nonignorable may be, the reasoning in this book is not significantly facilitated or clarified by a regular use of these terms. As a result, they are seldom seen here.

Another term often heard is 'unconfounded response mechanism'. It is related to (but not, in Rubin's terminology, exactly equal to) an ignorable response mechanism. Lee *et al.* (2001) call a response mechanism $q(r|s) = q(r|s, \mathbf{x}_s, y_s)$ unconfounded when $q(r|s, \mathbf{x}_s, y_s) = q(r|s, \mathbf{x}_s)$ and $\Pr(k \in r|s) > 0$

for every $k \in s$. Otherwise it is confounded. If the mechanism is ignorable it can depend on the sample auxiliary data $\mathbf{x}_s$ (that is, the data $\mathbf{x}_k$ for $k \in s$) but not on the sample study variable data y_s (the data y_k for $k \in s$).

Another concept in the literature is 'response propensity weighting'. The term is used in reference to a method described by Little (1988) as follows: 'The method is to regress the binary nonresponse indicator r on $\mathbf{x}$, using logistic or probit regression if necessary, and to derive predicted response propensities $\hat{p}_i = \Pr(r_i = 1|\mathbf{x}_i)$ for respondents and nonrespondents. Weights are then defined as proportional to the inverse propensities for respondents, or by forming adjustment cells based on the propensity score.'

The adjustment cells in question can be constructed by first forming intervals of the propensities $\hat{p}_i$ (corresponding to our $\hat{\theta}_k$) and then determining an adjustment weight, applied to all elements in one and the same interval, and defined as the inverse of the proportion of responding elements in the interval. If indeed the nonresponse occurs with the same probability within population groups with the same (or nearly the same) value of $\mathbf{x}$, we can expect the bias of the adjusted estimator to be small.

The method of response propensity weighting consists in essence of computing for $k \in s$ the estimated response probability $\hat{\theta}_k$, defined by the logit formula (8.16) or by a probit or other formula. For a fixed make-up of the auxiliary vector $\mathbf{x}_k$, there is a corresponding set of fitted values $\hat{\theta}_k$, computable for $k \in s$.

Formula (10.8) in Section 10.6 defines an indicator, $IND1\text{A}$, that can be used for measuring how alternative auxiliary vectors $\mathbf{x}_k$ succeed in predicting the unknown response probabilities. The greater the value of $IND1\text{A}$, the better the chances that weights based on the associated vector $\mathbf{x}_k$ will provide efficient protection against bias.

The calibration weights $w_k = d_k v_k$ given by (6.10), or more generally by (6.14), have embedded in them the notion of inverse response probability weighting. The value $d_k v_k$ can be seen as a substitute for $d_k \phi_k = d_k/\theta_k$. Therefore the weights $d_k v_k$ for $k \in r$ are useful for variance estimation, as Proposition 11.1 suggests.

APPENDIX: PROOF OF PROPOSITION 9.1

The proof of Proposition 9.1 proceeds as follows. For InfoUS, we have $\mathbf{x}_k = \begin{pmatrix} \mathbf{x}_k^* \\ \mathbf{x}_k^\circ \end{pmatrix}$ with a corresponding information input $\mathbf{X} = \begin{pmatrix} \sum_U \mathbf{x}_k^* \\ \sum_s d_k \mathbf{x}_k^\circ \end{pmatrix}$. InfoU and InfoS are the special cases obtained if $\mathbf{x}_k = \mathbf{x}_k^*$ and $\mathbf{x}_k = \mathbf{x}_k^\circ$, respectively. The calibration estimator with the weights defined by (6.14), using $d_{\alpha k} = d_k$ as initial weights, can be written as

$$\hat{Y}_{\text{W}} = \sum_r w_k y_k = \sum_r d_k v_k y_k = \sum_r d_k y_k + \left(\mathbf{X} - \sum_r d_k \mathbf{x}_k \right)' \mathbf{B}_{r;d} \quad (A9.1)$$

where

$$\mathbf{B}_{r;d} = \begin{pmatrix} \mathbf{B}_{r;d}^* \\ \mathbf{B}_{r;d}^\circ \end{pmatrix} = \left(\sum_r d_k \mathbf{z}_k \mathbf{x}_k' \right)^{-1} \sum_r d_k \mathbf{z}_k y_k. \quad (A9.2)$$

This vector of dimension $J^* + J^\circ$ is computable from the available data (except in the unlikely occurrence of a singularity) and is composed of two parts $\mathbf{B}^*_{r;d}$ and $\mathbf{B}^\circ_{r;d}$, corresponding to $\mathbf{x}^*_k$ and $\mathbf{x}^\circ_k$, respectively. Both $\mathbf{B}^*_{r;d}$ and $\mathbf{x}^*_k$ have dimension $J^* \geq 1$, while both $\mathbf{B}^\circ_{r;d}$ and $\mathbf{x}^\circ_k$ have dimension $J^\circ \geq 1$. We can view $\mathbf{B}_{r;d}$ as a regression coefficient (derived with instrument vector) for the regression of y_k on $\mathbf{x}_k$, weighted with the d_k. Two random sums appear in $\mathbf{B}_{r;d}$. Their pq-expected values are, respectively,

$$E_{pq}\left(\sum_r d_k \mathbf{z}_k \mathbf{x}'_k\right) = \sum_U \theta_k \mathbf{z}_k \mathbf{x}'_k, \qquad E_{pq}\left(\sum_r d_k \mathbf{z}_k y_k\right) = \sum_U \theta_k \mathbf{z}_k y_k.$$

For large response sets, $\mathbf{B}_{r;d}$ is close in probability to the population vector obtained by replacing the two sums in $\mathbf{B}_{r;d}$ with their respective expected values. This vector is $\mathbf{B}_{U;\theta}$ given by (9.15), which we can write as $\mathbf{B}_{U;\theta} = \begin{pmatrix} \mathbf{B}^*_{U;\theta} \\ \mathbf{B}^\circ_{U;\theta} \end{pmatrix}$, where $\mathbf{B}^*_{U;\theta}$ and $\mathbf{B}^\circ_{U;\theta}$ are the parts corresponding to $\mathbf{x}^*_k$ and $\mathbf{x}^\circ_k$. It is important to note that $\mathbf{B}_{U;\theta}$ is a constant, nonrandom vector. It is unknown, because of its dependence on θ_k and y_k for all $k \in U$. When population, sample and response set all increase, $\mathbf{B}_{r;d}$ tends to in probability to $\mathbf{B}_{U;\theta}$. That is,

$$\text{plim } \mathbf{B}_{r;d} = \text{plim}\left\{ \left(\sum_r d_k \mathbf{z}_k \mathbf{x}'_k\right)^{-1} \left(\sum_r d_k \mathbf{z}_k y_k\right) \right\}$$

$$= \left(\sum_U \theta_k \mathbf{z}_k \mathbf{x}'_k\right)^{-1} \left(\sum_U \theta_k \mathbf{z}_k y_k\right) = \mathbf{B}_{U;\theta}.$$

The difference $\mathbf{B}_{r;d} - \mathbf{B}_{U;\theta}$, although not exactly $\mathbf{0}$, is near $\mathbf{0}$. More importantly, it gets closer to $\mathbf{0}$ with high probability as the response set grows larger. Under general conditions, $\mathbf{B}_{r;d} - \mathbf{B}_{U;\theta}$ is of order in probability $O_p(m^{-1/2})$, where m is the size of r. Now, in (A9.1) we use $\mathbf{B}_{r;d} = \mathbf{B}_{U;\theta} + (\mathbf{B}_{r;d} - \mathbf{B}_{U;\theta})$ to eventually obtain

$$\hat{Y}_W = \sum_r d_k y_k + \left(\mathbf{X} - \sum_r d_k \mathbf{x}_k\right)' \mathbf{B}_{U;\theta} + \left(\mathbf{X} - \sum_r d_k \mathbf{x}_k\right)' (\mathbf{B}_{r;d} - \mathbf{B}_{U;\theta}).$$

$$\text{(A9.3)}$$

The transition from (A9.1) to (A9.3) is thus a result of centring the random vector $\mathbf{B}_{r;d}$ on its nonrandom counterpart $\mathbf{B}_{U;\theta}$. This creates a term, $\left(\mathbf{X} - \sum_r d_k \mathbf{x}_k\right)' (\mathbf{B}_{r;d} - \mathbf{B}_{U;\theta})$, which is of lower order of importance, compared to the preceding two terms on the right-hand side of (A9.3).

Now use the fact that $\mathbf{X}'\mathbf{B}_{U;\theta} = \left(\sum_U \mathbf{x}^*_k\right)' \mathbf{B}^*_{U;\theta} + \left(\sum_s d_k \mathbf{x}^\circ_k\right)' \mathbf{B}^\circ_{U;\theta}$ and $\left(\sum_r d_k \mathbf{x}_k\right)' \mathbf{B}_{U;\theta} = \left(\sum_r d_k \mathbf{x}^*_k\right)' \mathbf{B}^*_{U;\theta} + \left(\sum_r d_k \mathbf{x}^\circ_k\right)' \mathbf{B}^\circ_{U;\theta}$. From (A9.3) subtract the target parameter value $Y = \sum_U y_k$ and use $e_{\theta k} = y_k - \mathbf{x}'_k \mathbf{B}_{U;\theta}$ as defined by (9.15). We can then write the error of $\hat{Y}_W$ as

$$\hat{Y}_W - Y = \sum_r d_k e_{\theta k} - \sum_U e_{\theta k} + \left(\sum_s d_k \mathbf{x}^\circ_k - \sum_U \mathbf{x}^\circ_k\right)' \mathbf{B}^\circ_{U;\theta}$$

$$- \left(\sum_r d_k \mathbf{x}_k - \mathbf{X}\right)' (\mathbf{B}_{r;d} - \mathbf{B}_{U;\theta}).$$

To evaluate the bias of $\hat{Y}_W$, we take the pq-expectation term by term on the right-hand side. The pq-expectation of $\sum_r d_k e_{\theta k} - \sum_U e_{\theta k}$ is $-\sum_U (1 - \theta_k) e_{\theta k}$

and that of $\left(\sum_s d_k \mathbf{x}_k^\circ - \sum_U \mathbf{x}_k^\circ\right)' \mathbf{B}_{U;\theta}^\circ$ is 0, because $\mathbf{B}_{U;\theta}^\circ$ is constant, and $\sum_s d_k \mathbf{x}_k^\circ$ is unbiased for $\sum_U \mathbf{x}_k^\circ$. Hence,

$$B_{pq}(\hat{Y}_W) = E(\hat{Y}_W) - Y = \text{nearbias}(\hat{Y}_W) - R \qquad (A9.4)$$

where $\text{nearbias}(\hat{Y}_W)$ is given by (9.14) and the rest is

$$R = E_{pq}\left[\left(\sum_r d_k \mathbf{x}_k - \mathbf{X}\right)' \left(\mathbf{B}_{r;d} - \mathbf{B}_{U;\theta}\right)\right].$$

Although R is not exactly zero, it is near zero, because $(\mathbf{B}_{r;d} - \mathbf{B}_{U;\theta})$ tends to zero in probability with increasing size of the response set. Therefore,

$$B_{pq}(\hat{Y}_W) \approx \text{nearbias}(\hat{Y}_W)$$

as claimed by Proposition 9.1. The proof is therefore complete.

We note that $\text{nearbias}(\hat{Y}_W)$ is (i) independent of the sampling design $p(s)$ and (ii) the same for $\mathbf{x}_k = \mathbf{z}_k = \mathbf{x}_k^* = \mathbf{x}_{0k}$ (InfoU) as it is for $\mathbf{x}_k = \mathbf{z}_k = \mathbf{x}_k^\circ = \mathbf{x}_{0k}$ (InfoS), where $\mathbf{x}_{0k}$ is some fixed vector of auxiliary variables.

Note, however, that the *exact* bias, given by (A9.4), depends, through the usually negligible term R, on $p(s)$ and on whether an auxiliary variable carries information up to U (so that it belongs in $\mathbf{x}_k^*$) or only up to s (so that it belongs in $\mathbf{x}_k^\circ$).

CHAPTER 10

Selecting the Most Relevant Auxiliary Information

10.1. DISCUSSION

In a number of countries, administrative registers provide extensive sources of auxiliary information. As for registers on individuals, the situation is particularly favourable in the Scandinavian countries, and reliance on high-quality registers on individuals is increasing in other countries, particularly in the European Union. Complete (or very nearly complete) registers can give access to variables such as address, sex, age, income, civil status and country of birth. A variety of additional variables, relating to areas such as occupation, education and so on, may come from matching with other registers. Altogether, these variables provide a rich source of information.

Business surveys rely, in many countries, on well-developed business registers, supplying auxiliary information for surveys directed to the economic sector.

This begs the question as to how one should go about selecting the most relevant part of the total available information. Not all of it may necessarily be relevant.

Auxiliary information can be used both at the design stage (in constructing the sampling design) and at the estimation stage (in constructing the estimators). Here we concentrate on the second type of usage. The available register variables, except for particularly sensitive ones, may be used in the nonresponse analysis, in the computation of calibrated weights and in the construction of imputed values.

Chapters 6–8 have shown that calibration offers a flexible approach to the use of auxiliary information. There are very few restrictions on the auxiliary vector $\mathbf{x}_k$ used in the computation of the calibrated weights. Even large vectors can be handled by existing software. However, even if technically feasible, the use of all available auxiliary variables may not necessarily be the preferred solution. Judgement must be exercised in selecting the auxiliary information that will finally be retained in the estimation.

An inspection of the set of calibrated weights is always recommended. Excessively large weights should not be accepted. Many users prefer not to allow negative weights. Controls should therefore be built into the weight computation

Estimation in Surveys with Nonresponse   C.-E. Särndal and S. Lundström
© 2005 John Wiley & Sons, Ltd

to avoid weights outside a specified range. Some software programs have such controls built in.

The accuracy of an estimate is determined by the MSE, which is the sum of the variance and the squared bias. If the nonresponse is considerable and not counteracted by effective adjustment, the squared bias term is likely to dominate the MSE. This is the case especially if the sample size is large and, consequently, the variance small. A sizeable bias will reduce the possibilities for correct statistical inference. Valid confidence intervals cannot be computed, and the survey results lose much of their value.

The objective of overriding importance is therefore to reduce the bias as far as possible. One must identify auxiliary variables that meet this objective. The task is not simple, because it is usually impossible to assess or estimate the bias. Subjective judgements become to some extent necessary. Nevertheless, it is important to have a structured approach for 'selecting the best' from a larger set of potential auxiliary variables. We outline such an approach in this chapter. Useful guidelines are formulated for the choice of auxiliary variables. Computable indicators are proposed as tools for constructing a good auxiliary vector.

The chapter ends with a literature review. The choice of auxiliary information has been examined from different angles in the literature. A literature search on the use of auxiliary information reveals two kinds of articles: on the one hand, those that emphasize a reduction of the sampling variance; and on the other hand, those that emphasize bias reduction. Some articles address both aspects jointly.

10.2. GUIDELINES FOR THE CONSTRUCTION OF AN AUXILIARY VECTOR

The search for an efficient auxiliary vector should be guided by the following principles, which were mentioned informally in Section 3.3.

Principle 1

The auxiliary vector (or the instrument vector, if different from the auxiliary vector) should explain the inverse response probability, called the response influence.

□

Principle 2

The auxiliary vector should explain the main study variables. □

Principle 3

The auxiliary vector should identify the most important domains. □

The principal justification for the first two principles comes from Propositions 9.1–9.3. Principle 1 is also supported by results, derived for related procedures and conditions, in Fuller *et al.* (1994), and Principle 2 essentially confirms conclusions in Bethlehem (1988).

When Principle 1 is fulfilled, the bias is reduced in the calibration estimates for *all* study variables. This generality is important, because a large survey will ordinarily have many study variables, and we need assurance of effective bias removal in *all* estimates. Thus Principle 1 is particularly relevant.

If Principle 2 is fulfilled, the bias is reduced in the estimates for the *main* study variables, but perhaps not in the estimates (made with the same weights) for other study variables. The main study variables in a survey can usually be identified, but on subjective grounds. When Principle 2 is satisfied, the variance of the estimates will also be reduced.

Example 4.6 illustrated the fact that the residuals that determine the variance are likely to be smaller if the auxiliary vector can be formulated to closely identify the principal domains. Small residuals for the domain are also desirable from the standpoint of reducing the bias. This is the motivation behind Principle 3.

In the following we examine in particular how Principles 1 and 2 help in reducing the bias of the calibration estimator $\hat{Y}_W$.

10.3.   THE PROSPECTS FOR NEAR-ZERO BIAS WITH TRADITIONAL ESTIMATORS

In Chapter 7 we derived several traditional estimators as special cases of the calibration estimator $\hat{Y}_W$. They correspond to various formulations of the auxiliary vector $\mathbf{x}_k$, from the very simplest one to more advanced (but still rather basic) ones. Here we revisit these examples in order to illustrate the tendency of the bias to decrease as the $\mathbf{x}_k$-vector becomes increasingly powerful.

The eight traditional estimators reviewed in Chapter 7 are: the expansion estimator (EXP), the population weighting adjustment estimator (PWA), the weighting class estimator (WC), the ratio estimator (RA), the regression estimator (REG), the separate ratio estimator (SEPRA), the separate regression estimator (SEPREG) and the two-way classification estimator (TWOWAY). The respective formulae are (7.1), (7.4), (7.6), (7.9), (7.12), (7.14), (7.17) and (7.20). They were obtained from the general formulation (6.14) of $\hat{Y}_W$ when $d_{\alpha k} = d_k$, and the auxiliary vector $\mathbf{x}_k$ and the instrument vector $\mathbf{z}_k$ are as specified in Table 10.1. Since PWA and WC share the same $\mathbf{x}_k$ and $\mathbf{z}_k$ (but have different information inputs), there are seven different lines in the table, where $\boldsymbol{\gamma}_k = (\gamma_{1k}, \ldots, \gamma_{pk}, \ldots, \gamma_{Pk})'$ denotes the one-way classification vector defined by (7.3), and $\boldsymbol{\delta}_k = (\delta_{1k}, \ldots, \delta_{hk}, \ldots, \delta_{H-1,k})'$, as explained in Section 7.6.

Let us examine how well the eight traditional estimators succeed in satisfying the first two principles in Section 10.2. We start with Principle 1: the auxiliary vector (or the instrument vector) should explain the response influence.

For every case in Table 10.1, the condition $\boldsymbol{\mu}'\mathbf{z}_k = 1$ holds for some constant vector $\boldsymbol{\mu}$ and for all k. For example, when $\mathbf{z}_k = \mathbf{x}_k = \boldsymbol{\gamma}_k = (\gamma_{1k}, \ldots, \gamma_{pk}, \ldots, \gamma_{Pk})'$, the constant vector $\boldsymbol{\mu} = (1, 1, \ldots, 1)'$ satisfies the condition. Then Proposition 9.2 states that the nearbias is zero if the response influence $\phi_k = 1/\theta_k$ has the form $\phi_k = \boldsymbol{\lambda}'\mathbf{z}_k$. Using the seven expressions given for $\mathbf{z}_k$ in Table 10.1, we arrive at Table 10.2.

Table 10.1 Specifications of the auxiliary vector $\mathbf{x}_k$ and the instrument vector $\mathbf{z}_k$ leading to eight traditional estimators. Notation is as in Chapter 7.

Estimator	Auxiliary vector $\mathbf{x}_k$	Instrument $\mathbf{z}_k$
EXP	1	1
WC and PWA	$\boldsymbol{\gamma}_k$	$\boldsymbol{\gamma}_k$
RA	x_k	1
SEPRA	$x_k\boldsymbol{\gamma}_k$	$\boldsymbol{\gamma}_k$
REG	$(1, x_k)'$	$(1, x_k)'$
SEPREG	$(\boldsymbol{\gamma}_k', x_k\boldsymbol{\gamma}_k')'$	$(\boldsymbol{\gamma}_k', x_k\boldsymbol{\gamma}_k')'$
TWOWAY	$(\boldsymbol{\gamma}_k', \boldsymbol{\delta}_k')'$	$(\boldsymbol{\gamma}_k', \boldsymbol{\delta}_k')'$

Table 10.2 Relationship between the influence ϕ_k and the instrument vector $\mathbf{z}_k$ (as shown in Table 10.1) leading to zero nearbias for eight traditional estimators. a, a_p, b, b_p and b_h denote constants.

Estimator	Form of ϕ_k leading to zero nearbias	Description of the ϕ_k that lead to zero nearbias
EXP	$\phi_k = a$ for all $k \in U$	constant throughout
WC and PWA	$\phi_k = a_p$ for all $k \in U_p$	constant within groups
RA	$\phi_k = a$ for all $k \in U$	constant throughout
SEPRA	$\phi_k = a_p$ for all $k \in U_p$	constant within groups
REG	$\phi_k = a + bx_k$	linear in x_k
SEPREG	$\phi_k = a_p + b_px_k$	linear in x_k within groups
TWOWAY	$\phi_k = a_p + b_h$	additive without interaction

The mental image that we should refer to is the following. Imagine that we can ascertain the value of θ_k, and thus $\phi_k = 1/\theta_k$, for every $k \in U$, and that we can inspect the N points $(\phi_k, \mathbf{z}_k)$ for $k = 1, 2, \ldots, N$. (In reality, the ϕ_k are unknown, so the exercise is purely hypothetical.) If the relationship is perfect, so that $\phi_k = \boldsymbol{\lambda}'\mathbf{z}_k$ holds for every $k \in U$ and some vector $\boldsymbol{\lambda}$, then the bias of the estimator is virtually eliminated. If the relationship is close to perfect, so that $\phi_k \approx \boldsymbol{\lambda}'\mathbf{z}_k$, the bias is small. The required perfect linear relationship is stated in the second column of Table 10.2, and is further described in the third column.

We can now ask which of the eight estimators in Table 10.2 is likely to come closest to meeting the objective of a zero nearbias. The table states that the nearbias for the EXP and RA estimator is zero if the response weights are constant throughout the whole population. This is highly unlikely. Many studies on survey participation have shown that the response probability of individuals varies with observable factors such as age, sex and others. The prospects are brighter when a grouping is involved, as for the PWA, WC and SEPRA estimators. It is then sufficient that the ϕ_k, and hence the θ_k, be constant for all elements within a group.

When the auxiliary variables are derived from a register comprising the entire target population, x_k is a known value for all $k \in U$, and we can calculate both the auxiliary population total $\sum_U x_k$, required for the RA estimator, and the population count N. The REG estimator requires both $\sum_U x_k$ and N. Table 10.2 shows that the REG estimator has a zero nearbias if $\phi_k = a + bx_k$ holds for all k and some constants a and b. This condition is more likely to hold approximately than $\phi_k = a$, which is the severe condition under which the RA estimator (and the EXP estimator) achieve a zero nearbias. This favours the use of the REG estimator rather than the RA estimator. The relation $\phi_k = a_p + b_p x_k$ for SEPREG has an even stronger potential for holding approximately, because each of the p groups may then have its own linear relationship between ϕ_k and x_k.

The TWOWAY estimator has zero nearbias if ϕ_k is structured in an additive way, as the sum of an effect $a_p, p = 1, \ldots, P$, for the first classification and an effect $b_h, h = 1, \ldots, H$, for the second classification. We conclude that among the eight alternatives in Table 10.2, SEPREG and TWOWAY have the best potential for achieving the objective of a nearly zero bias.

A complete crossing of the two classifications into $P \times H$ cells is a possibility not accounted for in Table 10.2. The calibration estimator for this situation has zero nearbias if ϕ_k has a constant value a_{ph} for all elements k in the cell ph, $p = 1, \ldots, P, h = 1, \ldots, H$. This condition is more likely to be satisfied than the requirement $\phi_k = a_p + b_h$ for TWOWAY.

Table 10.2 communicates the following message: the better we succeed in incorporating important auxiliary information into the auxiliary vector, the better are the prospects for a bias reduced to near-zero levels. In practice, we are usually not limited to the seven vector formulations in the table. In many surveys, a considerably more extensive pool of potential auxiliary variables is available, and the survey methodologist is in a position to extract from this pool those variables deemed to be the most valuable.

Remark 10.1

Table 10.2 prompts an interesting observation on the use of a quantitative variable x. Such a variable is used in the estimator RA. For example, in a business survey, x may be a size-related measure such as 'number of employees' for an enterprise. For such a variable, we have an option: we can use it in its original, quantitative form, as in the estimator RA, or we can use it to form a classification of the elements in the population or in the sample. By means of the known values x_k, for $k \in U$ or $k \in s$, we create size groups – for example, small, medium, or large elements, if only three groups are specified. Our discussion of Table 10.2 favours the option of forming a grouping from the quantitative variable values. The reason is that if an x-variable is used in its original continuous form, as in RA, then a nearly zero bias is achieved under the severe condition of a constant response probability throughout. This is less likely to hold than the more liberal 'constant within groups' condition needed when we use the x-variable to classify elements and use PWA. Although RA may have smaller variance than PWA, its bias may be greater.   □

Remark 10.2

Modelling the response behaviour is a prominent topic in the vast literature on nonresponse adjustment. There was no need to invoke models to arrive at Proposition 9.2. Nevertheless, the proposition offers a perspective on the statistician's penchant for relying on assumptions. If we truly believe in a response behaviour such that $\phi_k = 1 + \lambda'\mathbf{z}_k$ holds for every $k \in U$ and some constant vector λ, then Proposition 9.2 supports the conclusion that 'the calibration estimator $\hat{Y}_W$ is nearly unbiased'. In particular, if $\hat{Y}_W$ is built on $\mathbf{x}_k = \mathbf{z}_k = \gamma_k$, the one-way classification vector, then Table 10.2 says that $\hat{Y}_{WC}$ and $\hat{Y}_{PWA}$ are nearly without bias if the ϕ_k (and thus the θ_k) are constant within groups, as when nonresponse occurs at random within groups. This is also described as 'y-values are missing at random within groups'. But the survey conditions may allow $\hat{Y}_W$ to be constructed on a variety of alternative auxiliary vectors. Suppose $\mathbf{x}_k = \mathbf{x}_{0k}$ is the vector of our choice, and $\mathbf{z}_k = \mathbf{x}_{0k}$. If we decide to call the hypothetical relationship $\phi_k = 1 + \lambda'\mathbf{x}_{0k}$ 'our model', then '$\hat{Y}_W$ is nearly unbiased under the model' is a correct statement. Regardless of which one of the many possible $\mathbf{x}_k$-vectors we finally settle on, there is always an assumption that supports the reassuring statement 'despite nonresponse, this survey delivers nearly unbiased estimates'. It is true even under the most primitive specification, $\mathbf{x}_k = \mathbf{z}_k = 1$ for all k. No model is ever true, but some are more effective. Among the $\mathbf{x}_k$-vectors that can enter into consideration, some bring better bias protection than others, but a perfect one cannot be found.   □

   We turn to Principle 2: the auxiliary vector should explain the main study variables. It follows from Proposition 9.3 that the nearbias of $\hat{Y}_W$ is zero if $e_k = y_k - \mathbf{x}_k'\beta = 0$ for all k and some constant β, that is, if $\mathbf{x}_k$ fully explains y_k. Let us examine whether this condition is likely to hold for the eight estimators in Table 10.1.
   The thought process is now as follows. Suppose that we could examine the set of N points $\{(y_k, \tilde{y}_k) : k = 1, \ldots, N\}$, where $\tilde{y}_k = \mathbf{x}_k'\beta$. If the relationship is perfect, so that $\tilde{y}_k$ equals y_k for every k and some vector β, then the bias is nearly eliminated. For each estimator, this perfect (but in practice unattainable) relationship is stated in the second column of Table 10.3 and further explained in the third column.
   We can now ask which of the eight estimators in Table 10.3 is likely to succeed best in yielding small differences $y_k - \tilde{y}_k$ for all k, and thus a small bias. The bias for the EXP estimator will be small if y_k is nearly constant throughout the population. This zero-variance property is a far-fetched possibility. The condition for PWA and WC is that y_k be constant within groups. This stands a somewhat better chance of approximating the truth. The implication is that we should attempt to identify groups, which are as far as possible homogeneous with respect to the y-variable, a motivation akin to that used in constructing efficient strata for stratified sampling. The RA estimator is in a better position than the EXP estimator to give small residuals. Still better placed is the REG estimator, since it also allows the freedom of a nonzero intercept. The SEPRA and SEPREG estimators are favoured if each group has its own linear regression,

Table 10.3   The linear relationship between the study variable and the auxiliary vector leading to zero nearbias for eight well-known estimators. α, α_p, β, β_p and β_h denote constants.

Estimator	Form of y_k that yields zero nearbias	Description of the y_k that yield zero nearbias
EXP	$y_k = \alpha$ for all $k \in U$	constant throughout
WC and PWA	$y_k = \beta_p$ for all $k \in U_p$	constant within groups
RA	$y_k = \alpha x_k$ for all $k \in U$	linear in x_k through the origin
SEPRA	$y_k = \alpha_p x_k$ for all $k \in U_p$	linear in x_k through the origin within groups
REG	$y_k = \alpha + \beta x_k$	linear in x_k
SEPREG	$y_k = \alpha_p + \beta_p x_k$ for all $k \in U_p$	linear in x_k within groups
TWOWAY	$y_k = \alpha_p + \beta_h$	additive effects, no interaction

and TWOWAY is indicated if there are additive group effects. Of the alternatives in the table, SEPREG and TWOWAY show the best signs of achieving small residuals.

The message is clear: it is in the interest of reducing both bias and variance to seek out the best possible x-variables within the pool of potential variables, and to use them to build an $\mathbf{x}_k$-vector that gives even better chances of small residuals than any alternative in Table 10.3.

Remark 10.3

If called for, a quantitative (and positive) auxiliary variable may be transformed. Instead of x_k, we may prefer to work with $\sqrt{x_k}$ or $\ln x_k$ or some other transformed value. If x_k is specified in the frame for every $k \in U$, we can compute $\sqrt{x_k}$ for every $k \in U$ and thus the auxiliary total $\sum_U \sqrt{x_k}$. By contrast, if the population total must be imported, the transformed variable may not be a functional auxiliary variable: if $\sum_U x_k$ can be obtained from an outside source, it does not follow that $\sum_U \sqrt{x_k}$ can be obtained.   □

10.4.   FURTHER AVENUES TOWARDS A ZERO BIAS

Formula (9.18) offers further insights into the conditions for $\hat{Y}_W$ to achieve a nearly zero bias. We assume the standard specification $\mathbf{z}_k = \mathbf{x}_k$, such that $\boldsymbol{\mu}'\mathbf{x}_k = 1$ holds for all k. By (9.18) the nearbias is zero, regardless of the sampling design, if $\sum_U \theta_k \mathbf{x}_k e_k = 0$, where the residuals are $e_k = y_k - \mathbf{x}_k'\mathbf{B}_U$ with $\mathbf{B}_U = \left(\sum_U \mathbf{x}_k \mathbf{x}_k'\right)^{-1} \sum_U \mathbf{x}_k y_k$. If $\mathbf{x}_k = (x_{1k}, \ldots, x_{jk}, \ldots, x_{Jk})'$, $\sum_U \theta_k \mathbf{x}_k e_k = 0$ can be written as

$$\sum_U \theta_k x_{jk} e_k = 0, \qquad j = 1, \ldots, J. \qquad (10.1)$$

If these J conditions hold simultaneously, the nearbias is zero. Let us examine a few simple auxiliary vectors to see the implications of (10.1).

Consider first the one-way classification, indicated by the vector $\boldsymbol{\gamma}_k = (\gamma_{1k}, \ldots, \gamma_{pk}, \ldots, \gamma_{Pk})'$. As seen in Section 7.3, the calibration estimator is $\hat{Y}_{\text{PWA}}$ if $\mathbf{x}_k = \mathbf{x}_k^* = \boldsymbol{\gamma}_k = \mathbf{z}_k$, and $\hat{Y}_{\text{WC}}$ if $\mathbf{x}_k = \mathbf{x}_k^\circ = \boldsymbol{\gamma}_k = \mathbf{z}_k$. The residuals are $e_k = y_k - \overline{y}_{U_p}$ for $k \in U_p$. The nearbias is same for $\hat{Y}_{\text{PWA}}$ and $\hat{Y}_{\text{WC}}$. By (10.1) with $J = P$, the nearbias is zero if P conditions hold simultaneously, namely,

$$\sum\nolimits_{U_p} \theta_k (y_k - \overline{y}_{U_p}) = 0, \qquad p = 1, \ldots, P.$$

That is, it is sufficient for nearly zero bias that θ_k and y_k are uncorrelated in every group U_p. Although this is of some theoretical interest, it is difficult or impossible in practice to know whether or not the P conditions are satisfied. They are indeed satisfied if (but not only if) the response probability is constant group by group, as Table 10.2 has already pointed out.

Consider next the auxiliary vector $\mathbf{x}_k = \mathbf{x}_k^* = (1, x_k)'$ discussed in Section 7.4. The information input is $\mathbf{X} = \sum_U \mathbf{x}_k = (N, \sum_U x_k)'$. The residuals are $e_k = y_k - \overline{y}_U - B_U(x_k - \overline{x}_U)$ with $B_U = \left[\sum_U (x_k - \overline{x}_U) y_k\right] / \left[\sum_U (x_k - \overline{x}_U)^2\right]$. It follows from (10.1) that the nearbias of the calibration estimator for this auxiliary vector, $\hat{Y}_{\text{W}} = \hat{Y}_{\text{REG}}$, is zero if two conditions hold: (i) $\sum_U \theta_k e_k = 0$ and (ii) $\sum_U \theta_k x_k e_k = 0$. The first condition is equivalent to $r_{\theta y} = r_{\theta x} r_{xy}$, where $r_{\theta y} = S_{\theta y U}/S_{\theta U} S_{y U}$ is the coefficient of correlation between θ and y in population U and $r_{\theta x}$ and r_{xy} are analogously defined correlation coefficients. The partial correlation coefficient between θ and y, controlling for x, is given by $r_{\theta y \cdot x} = (r_{\theta y} - r_{\theta x} r_{xy})/\{(1 - r_{\theta x}^2)(1 - r_{xy}^2)\}^{1/2}$. An educated guess might be that a partial correlation $r_{\theta y \cdot x} = 0$ would suffice to give a zero nearbias. It is in fact not sufficient, but if in addition $\sum_U \theta_k x_k e_k = 0$ holds, then $\hat{Y}_{\text{W}}$ is nearly unbiased. In practice it may be difficult to satisfy these two conditions. A more fruitful approach to controlling the bias may be, as pointed out earlier, to use the continuous variable values x_k to form size groups and then to compute $\hat{Y}_{\text{PWA}}$ or $\hat{Y}_{\text{WC}}$. Alternatively, we can use the known values x_k both to form size groups and as continuous measurements within groups, and then compute $\hat{Y}_{\text{SEPRA}}$ or $\hat{Y}_{\text{SEPREG}}$.

As a third example, we consider the two-way classification discussed in Section 7.6. For InfoU, the auxiliary vector is $\mathbf{x}_k = \mathbf{x}_k^*$ given by (7.19). The information input consists of the known population counts in the marginal groups $U_{p\bullet}$, $p = 1, \ldots, P$, and $U_{\bullet h}$, $h = 1, \ldots, H$. The counts in the cross-classification cells U_{ph} need not be known. The residuals $e_k = y_k - \mathbf{x}_k' \mathbf{B}_U$ have the form $e_k = y_k - B_p - B_h$ for $k \in U_{ph}$. In this case, (10.1) implies that the nearbias is zero if the θ_k are constant in every marginal group $U_{p\bullet}$, $p = 1, \ldots, P$, *and* also in every marginal group $U_{\bullet h}$, $h = 1, \ldots, H - 1$. A constant response probability within each cross-classification group U_{ph} does not suffice to achieve a zero nearbias, for that auxiliary vector.

10.5. A FURTHER TOOL FOR REDUCING THE BIAS

Two choices influence the bias of $\hat{Y}_{\text{W}}$, namely, the auxiliary vector $\mathbf{x}_k$ and the instrument vector $\mathbf{z}_k$. For an already fixed auxiliary vector $\mathbf{x}_k$, we have the freedom to choose the vector $\mathbf{z}_k$. If $\mathbf{z}_k$ is of the form $\mathbf{z}_k = c_k \mathbf{x}_k$, we have the freedom to choose the constants c_k. Thus the choice of $\mathbf{z}_k$ becomes a further tool for reducing the bias, after having fixed an auxiliary vector $\mathbf{x}_k$.

By way of illustration, let us see how the formulation of $\mathbf{z}_k$ may influence the bias when the auxiliary vector is univariate, $\mathbf{x}_k = x_k$.

As Table 10.1 shows, the RA estimator is tied to the specification $\mathbf{z}_k = 1$ for all k. The nearbias is zero if the response probabilities are constant throughout the population. If we have strong reason to believe this, then $\mathbf{z}_k = 1$ is an appropriate choice.

However, consider the more general specification $\mathbf{z}_k = z_k = x_k^{1-v}$, for a specified constant v. Inserting $z_k = x_k^{1-v}$ into (9.19), we conclude that the nearbias is zero if $\phi_k = 1 + a\, x_k^{1-v}$, where a is a constant. If there is reason to believe that the influence ϕ_k increases with x_k, we should choose a value v less than one. Conversely, if we believe that the influence ϕ_k is decreasing as x_k increases, we should choose a value of v greater than one.

A judgement is required here about the rate $1 - v$ at which the influence increases or decreases with the continuous variable x. It is rare in most surveys to have a sufficient understanding of the nonresponse mechanism to be able to specify v. As pointed out earlier, a usually better, more reliable use of the quantitative variable is to group the elements into a number of classes based on the size of x, and none of them exceedingly small.

10.6. THE SEARCH FOR A POWERFUL AUXILIARY VECTOR

To accommodate new nonresponse treatment procedures in a survey may require an investment of time and resources. The survey methodologist provides important input into this process.

The more one can succeed in incorporating relevant auxiliary variables into the $\mathbf{x}_k$-vector, the better, generally speaking, are the chances of a low bias in the calibration estimator $\hat{Y}_W$. Building a potent $\mathbf{x}_k$-vector is of paramount importance. The statistician must first make an inventory of potential auxiliary variables, then select the most pertinent variables. Principles 1–3 stated in Section 10.2 should guide this effort.

An important part of this construction process is to evaluate and compare alternative auxiliary vectors. This section presents computational tools, called indicators, that are helpful in satisfying Principles 1 and 2.

Indicator for Principle 1

A dilemma for estimation in the presence of nonresponse is that the influences $\phi_k = 1/\theta_k$ are unknown. For example, we cannot compute the unbiased estimator $\hat{Y} = \sum_r d_k \phi_k y_k$ of $Y = \sum_U y_k$. But we can compute the calibration estimator $\hat{Y}_W$ because its weights, given by (6.10) or (6.14), do not depend on the unknown ϕ_k. By Proposition 9.2, $\hat{Y}_W$ is nearly unbiased if ϕ_k is linearly related to the instrument $\mathbf{z}_k$ so that $\phi_k = 1 + \boldsymbol{\lambda}'\mathbf{z}_k$ holds for $k \in U$ and some constant, nonrandom vector $\boldsymbol{\lambda}$.

It is not possible to determine such a vector from the data available. But we can determine *proxies* $\hat{\phi}_k$ for the ϕ_k such that (i) $\hat{\phi}_k$ is linearly related to $\mathbf{z}_k$ for the specific realizations r and s, and (ii) the $\hat{\phi}_k$ give a correct calibration on $\mathbf{x}_k$ from r to s. These two requirements, $\hat{\phi}_k = 1 + \boldsymbol{\lambda}'_r \mathbf{z}_k$ for $k \in r$ and $\sum_r d_k \hat{\phi}_k \mathbf{x}_k = \sum_s d_k \mathbf{x}_k$, are satisfied by taking $\hat{\phi}_k = v_{sk}$, where

$$v_{sk} = 1 + \boldsymbol{\lambda}'_r \mathbf{z}_k = 1 + \left(\mathbf{X}_s - \sum_r d_k \mathbf{x}_k\right)' \left(\sum_r d_k \mathbf{z}_k \mathbf{x}'_k\right)^{-1} \mathbf{z}_k \qquad (10.2)$$

with

$$\mathbf{x}_k = \begin{pmatrix} \mathbf{x}^*_k \\ \mathbf{x}^\circ_k \end{pmatrix}$$

and

$$\mathbf{X}_s = \begin{pmatrix} \hat{\mathbf{X}}^* \\ \hat{\mathbf{X}}^\circ \end{pmatrix} = \begin{pmatrix} \sum_s d_k \mathbf{x}^*_k \\ \sum_s d_k \mathbf{x}^\circ_k \end{pmatrix}.$$

The information requirement for (10.2) is that $\mathbf{x}^*_k$ and $\mathbf{x}^\circ_k$ are known vector values for every $k \in s$. The proxies $\hat{\phi}_k = v_{sk}$ reflect the relation between the set of respondents, r, and the sample of elements, s. The v_{sk} mirror the response pattern that was actually obtained in the sample s.

By definition, the moon vector value $\mathbf{x}^\circ_k$ is known for every $k \in s$. When a population list or frame specifies the star vector $\mathbf{x}^*_k$ for every $k \in U$, it follows that $\mathbf{x}^*_k$ is also known for every $k \in s$, as discussed in Section 6.2, and (10.2) is readily computable. This holds, for example, for surveys of individuals, households or enterprises that rely on a total population register.

When $\mathbf{X}^* = \sum_U \mathbf{x}^*_k$ is an imported total, then $\mathbf{x}^*_k$ may not be individually known for every $k \in s$, and the component $\hat{\mathbf{X}}^*$ of $\mathbf{X}_s$ in (10.2) cannot be computed. We comment on this situation in Remark 10.4.

We want the proxies to reflect the individual differences existing among the sampled elements. The proxies (10.2) meet this objective in that $\hat{\phi}_k = v_{sk}$ is a function of the data we have about an element $k \in s$, namely, the couple $(\mathbf{x}_k, \mathbf{z}_k)$. The more the proxies vary, the better the prospects that $\hat{Y}_W$ will be protected against bias. We therefore use the (d-weighted) variance of $\hat{\phi}_k = v_{sk}$ given by (10.2) to indicate the strength of $\mathbf{x}_k$. This is given by

$$IND1 = \frac{1}{\sum_r d_k} \sum_r d_k (v_{sk} - \bar{v}_{s;r;d})^2 \qquad (10.3)$$

where $\bar{v}_{s;r;d} = \sum_r d_k v_{sk} / \sum_r d_k$ is the (d-weighted) mean of the v_{sk}. For complete response ($r = s$), we have $IND1 = 0$, as is reasonable. The general tendency is that the higher the nonresponse rate, the higher the value of $IND1$.

When $\mathbf{z}_k = \mathbf{x}_k$ and $\boldsymbol{\mu}'\mathbf{x}_k = 1$ for all k and some constant $\boldsymbol{\mu}$, then $IND1$ can be written as

$$IND1 = \frac{\left(\sum_s d_k \mathbf{x}_k\right)' \left(\sum_r d_k \mathbf{x}_k \mathbf{x}'_k\right)^{-1} \left(\sum_s d_k \mathbf{x}_k\right)}{\left(\sum_r d_k\right)} - \frac{\left(\sum_s d_k\right)^2}{\left(\sum_r d_k\right)^2}.$$

Under these conditions, we have $\bar{v}_{s:r;d} = \sum_s d_k / \sum_r d_k$, which depends not on $\mathbf{x}_k$, only on the sampling design and on the sets s and r.

The most primitive specifications, $\mathbf{x}_k = 1 = \mathbf{z}_k$, which yield the EXP estimator, give $IND1 = 0$. We also have $IND1 = 0$ when $\mathbf{x}_k = x_k$ and $\mathbf{z}_k = 1$ for all k, as for the RA estimator (7.9). This is in line with the earlier finding that RA is not well protected against nonresponse bias.

As the $\mathbf{x}_k$-vector expands through the addition of more and more x-variables, then the calibration restriction $\sum_r d_k \hat{\phi}_k \mathbf{x}_k = \sum_s d_k \mathbf{x}_k$ imposes more and more conditions on the proxies $\hat{\phi}_k = v_{sk}$ for $k \in r$. The variance of v_{sk} will tend to increase as more x-variables are added. The bias $B_{pq}(\hat{Y}_W)$ can be expected to decrease accordingly.

It can be an important step in the building of a powerful auxiliary vector to compute $IND1$ for alternative $\mathbf{x}_k$-vectors, and monitor the increasing tendency of $IND1$ as further x-variables, categorical or continuous, are added to $\mathbf{x}_k$. We illustrate this process by means of empirical examples in Section 10.7.

To illustrate $IND1$, consider the one-way classification, where $\mathbf{x}_k = \mathbf{z}_k = \boldsymbol{\gamma}_k = (\gamma_{1k}, \ldots, \gamma_{pk}, \ldots, \gamma_{Pk})'$ and $\mathbf{x}_k$ is known for every $k \in s$. In this case,

$$IND1 = \frac{1}{\sum_r d_k} \left(\sum_{p=1}^{P} \frac{\left(\sum_{s_p} d_k \right)^2}{\sum_{r_p} d_k} - \frac{\left(\sum_s d_k \right)^2}{\sum_r d_k} \right). \qquad (10.4)$$

In particular, for SI with $d_k = N/n$ for all k, (10.4) takes the still more readily interpreted form

$$IND1 = \frac{1}{m} \left(\sum_{p=1}^{P} \frac{n_p^2}{m_p} - \frac{n^2}{m} \right) \qquad (10.5)$$

where m_p is the number of responding elements out of n_p in group p, and m is the total number of respondents out of the n sampled elements.

It is obvious that $IND1$ can be made artificially large in cases where a chosen sample can be split into smaller and smaller groups, so that some or all m_p become very small. This is not the spirit in which the indicator should be used. The m_p are subject to considerable random fluctuations for small groups; all of them should be kept larger than a critical value. Whether that critical value should be 10, 20 or even higher depends on the survey conditions, and professional judgement should be used. A grouping based on, for example, age or a similarly continuous concept should not be overextended to allow more classes than the underlying concept can reasonably support. Instead, the objective is to monitor the development (the increase) of $IND1$ as the $\mathbf{x}_k$-vector expands by adding further x-variables, qualitative or quantitative. If two or more categorical concepts are involved, excessively small counts m_p are avoided by constructing the $\mathbf{x}_k$-vector in the manner of (7.19). That is, rather than to cross-classifying, one should string out the classifications 'side by side' and calibrate on the marginal information, population-based or sample-based.

Remark 10.4

When the total $\mathbf{X}^* = \sum_U \mathbf{x}_k^*$ is imported, and $\mathbf{x}_k^*$ is not individually known for every $k \in s$, we cannot (unless $\mathbf{x}_k = \mathbf{x}_k^\circ$) compute v_{sk} given by (10.2). In this case, the definition (10.3) of $IND\,1$ is not applicable. But we can compute

$$v_k = 1 + (\mathbf{X} - \sum_r d_k \mathbf{x}_k)' \left(\sum_r d_k \mathbf{z}_k \mathbf{x}_k'\right)^{-1} \mathbf{z}_k \text{ with } \mathbf{x}_k = \begin{pmatrix} \mathbf{x}_k^* \\ \mathbf{x}_k^\circ \end{pmatrix} \text{ and } \mathbf{X} = \begin{pmatrix} \mathbf{X}^* \\ \hat{\mathbf{X}}^\circ \end{pmatrix} =$$

$\begin{pmatrix} \sum_U \mathbf{x}_k^* \\ \sum_s d_k \mathbf{x}_k^\circ \end{pmatrix}$. For this situation we can instead define $IND\,1$ as the (d-weighted) variance of the v_k. That is, v_{sk} is replaced by v_k in (10.3). When both alternatives are open, it is preferable to use $IND\,1$ based on the v_{sk}. One reason is that the computation with v_k requires a preliminary computation of a number of population totals, which may be impractical when a number of alternative $\mathbf{x}_k$-vectors are to be compared. When v_k is used rather than v_{sk}, $IND\,1$ tends to take greater values; it will still serve as a useful tool for ranking alternative $\mathbf{x}_k$-vectors. □

Alternative indicator for Principle 1

To explain in the best possible manner the response probability θ_k in terms of the auxiliary vector $\mathbf{x}_k$ is the principal motivation behind the first step of the two-step calibration approach, as discussed in Chapters 8 and 9. The following alternative indicator capitalizes on this idea. If the θ_k were known, their variability in the population U would be measured by

$$SST = \sum_U (\theta_k - \overline{\theta}_U)^2 \tag{10.6}$$

where $\overline{\theta}_U = (1/N) \sum_U \theta_k$ is the mean response probability. Let $\tilde{\theta}_k$ be a value obtained for $k \in U$ by a fitting procedure that yields $\tilde{\theta}_k$ as a function of the individual auxiliary vector value $\mathbf{x}_k$. Then

$$SSR = \sum_U (\theta_k - \tilde{\theta}_k)^2 \tag{10.7}$$

is a measure of residual variance, and $1 - SSR/SST$ is a measure of variance explained, in the spirit of the coefficient of determination in regression analysis.

But the θ_k are unknown. We observe 'response or not' for the elements in the sample; the associated random variable takes the value $R_k = 1$ for $k \in r$ and $R_k = 0$ for $k \in s - r$. Its conditional expectation under the response distribution is $E_q(R_k|s) = \theta_k$. We replace θ_k, $\overline{\theta}_U$ and $\tilde{\theta}_k$ in (10.6) and (10.7) by the sample-based analogues R_k, $\overline{R}_{s;d}$ and $\hat{\theta}_k$, respectively, and obtain estimates of SST and SSR given, respectively, by

$$\sum_s d_k (R_k - \overline{R}_{s;d})^2 \quad \text{and} \quad \sum_s d_k (R_k - \hat{\theta}_k)^2$$

where $\overline{R}_{s;d} = \sum_s d_k R_k / \sum_s d_k$ and $\hat{\theta}_k = \mathbf{x}_k' \left(\sum_s d_k \mathbf{z}_k \mathbf{x}_k'\right)^{-1} \sum_s d_k \mathbf{z}_k R_k$ with $\mathbf{x}_k = \begin{pmatrix} \mathbf{x}_k^* \\ \mathbf{x}_k^\circ \end{pmatrix}$. Here, $\hat{\theta}_k$ is the least-squares prediction (with instrument vector) of θ_k. This can work well in many situations but is not ideal, because a small number of erratic predictions $\hat{\theta}_k$ may occur outside the unit interval. Instead, one may use the logit predictions given by (8.16) to ensure values $\hat{\theta}_k$ situated in the unit interval. The sample-based, computable measure of explained variation is

$$IND\,1\text{A} = 1 - \frac{\sum_s d_k (R_k - \hat{\theta}_k)^2}{\sum_s d_k (R_k - \overline{R}_{s;d})^2}. \tag{10.8}$$

The information requirement for $IND\,1$A is that $\mathbf{x}_k^*$ and $\mathbf{x}_k^{\circ}$ must be known for every $k \in s$. When this is satisfied, $IND\,1$A can be an alternative to $IND\,1$ as an indicator of the strength of the auxiliary vector $\mathbf{x}_k$.

When additional x-variables are included in the $\mathbf{x}_k$-vector, for fixed s and r, the value of $IND\,1$A will tend to increase. We can compute $IND\,1$A for different $\mathbf{x}_k$-vectors, and monitor its progression as $\mathbf{x}_k$ expands. The value of $IND\,1$A is likely to stay only modestly greater than zero, even for a powerful $\mathbf{x}_k$-vector, because of the dichotomous nature of the response variable R_k. Nevertheless, monitoring the progression of $IND\,1$A can give valuable guidance.

As an illustration, consider again the one-way classification, where $\mathbf{x}_k = \mathbf{z}_k = \boldsymbol{\gamma}_k = (\gamma_{1k}, \ldots, \gamma_{pk}, \ldots, \gamma_{Pk})'$. Then

$$IND\,1\text{A} = \frac{1}{T} \left(\sum_{p=1}^{P} \frac{\left(\sum_{r_p} d_k\right)^2}{\sum_{s_p} d_k} - \frac{\left(\sum_r d_k\right)^2}{\sum_s d_k} \right) \tag{10.9}$$

where

$$T = \left(\sum_r d_k\right) \left(1 - \frac{\sum_r d_k}{\sum_s d_k}\right).$$

Apart from the normalizing factor T in the denominator, (10.9) is the same expression as (9.11), derived with similar reasoning for the classification vector. It is also of interest to compare (10.9) with (10.4) obtained by $IND\,1$ for the same $\mathbf{x}_k$-vector. Although (10.9) and (10.4) are not equal, there is an obvious kinship.

In particular, for SI with $d_k = N/n$ for all k, (10.9) becomes

$$IND\,1\text{A} = \frac{1}{m(1 - m/n)} \left(\sum_{p=1}^{P} \frac{m_p^2}{n_p} - \frac{m^2}{n} \right) \tag{10.10}$$

which should be compared with the corresponding value of $IND\,1$ given by (10.5).

Faced with a choice between $IND\,1$ and $IND\,1$A, the former is recommended. It has greater stability.

Indicator for Principle 2

We know from Proposition 9.3 that the nearbias of $\hat{Y}_W$ is zero if the auxiliary vector explains the study variable, or, in particular, if the residuals $e_k = y_k - \mathbf{x}_k' \mathbf{B}_U$ are zero for all k, where $\mathbf{B}_U = \left(\sum_U \mathbf{z}_k \mathbf{x}_k'\right)^{-1} \sum_U \mathbf{z}_k y_k$. It is unrealistic to expect this to hold in practice, but if all residuals are small, the bias of $\hat{Y}_W$ will be small, and so will its variance. We therefore develop an indicator that measures how close the residuals are to zero.

The variability of the y_k in the population is given by the (unknown) quantity

$$SST_Y = \sum_U (y_k - \overline{y}_U)^2 \tag{10.11}$$

where $\bar{y}_U = (1/N) \sum_U y_k$ is the population mean. Let $\tilde{y}_k$ be a value obtained for $k \in U$ by a fitting procedure that delivers $\tilde{y}_k$ as a function of the individual auxiliary vector value $\mathbf{x}_k$. Then

$$SSR_Y = \sum_U (y_k - \tilde{y}_k)^2 \tag{10.12}$$

measures residual variance, and $(SST_Y - SSR_Y)/SST_Y$ is a measure of variance explained.

With the data we have, estimates of the unknown SST_Y and SSR_Y are given by, respectively,

$$\sum_r d_k v_{sk} (y_k - \bar{y}_{r;dv})^2 \quad \text{and} \quad \sum_r d_k v_{sk} (y_k - \hat{y}_k)^2$$

with

$$\bar{y}_{r;dv} = \frac{\sum_r d_k v_{sk} y_k}{\sum_r d_k v_{sk}}, \qquad \hat{y}_k = \mathbf{x}_k' \left(\sum_r d_k v_{sk} \mathbf{z}_k \mathbf{x}_k' \right)^{-1} \sum_r d_k v_{sk} \mathbf{z}_k y_k \tag{10.13}$$

where $\mathbf{x}_k = \begin{pmatrix} \mathbf{x}_k^* \\ \mathbf{x}_k^\circ \end{pmatrix}$ and v_{sk} is given by (10.2). The quantities $\bar{y}_{r;dv}$ and $\hat{y}_k$ capitalize on the idea, used earlier, that v_{sk} is a proxy for the unknown influence $\phi_k = 1/\theta_k$. A suggestion close in spirit to the predicted values $\hat{y}_k$ in (10.13) is found in Fuller *et al.* (1994). The indicator for the study variable y is

$$IND2 = 1 - \frac{\sum_r d_k v_{sk} (y_k - \hat{y}_k)^2}{\sum_r d_k v_{sk} (y_k - \bar{y}_{r;dv})^2}. \tag{10.14}$$

This indicator measures the capacity of the auxiliary vector $\mathbf{x}_k$ to explain the specific study variable y. Thus $IND2$ can have a comparatively high value for some study variables but not for others. The most primitive conditions, $\mathbf{x}_k = \mathbf{z}_k = 1$ for all k, give, not unexpectedly, $IND2 = 0$. As discussed in Remark 10.4, v_{sk} may have to be replaced in (10.14) by v_k when $\mathbf{x}_k^*$ is not individually known for every $k \in s$.

Remark 10.5

The computation of v_{sk} by formula (10.2) may yield a small number of undesirable values, which can affect the indicators $IND1$ and $IND2$. For example, the computation of $\hat{y}_k$ in (10.13) requires inversion of the matrix $\sum_r d_k v_{sk} \mathbf{z}_k \mathbf{x}_k'$. If $d_k v_{sk}$ happens to be negative for one or more large or otherwise influential elements k, the matrix may be singular or nearly singular, with erratic predictions $\hat{y}_k$ as a result. For survey practice, we recommend avoiding undesirable values by not allowing into the $\mathbf{x}_k$-vector any variables that are likely to cause undesirable weights. $\qquad \square$

10.7. EMPIRICAL ILLUSTRATIONS OF THE INDICATORS

The use of indicators can be an effective means of improving estimation with data affected by nonresponse. They assist the construction of an auxiliary vector that is likely to keep the bias within limits. The idea has been used before. For example, Bethlehem and Schouten (2004) examine an indicator which, like $IND2$ given by (10.14), is study variable dependent.

The indicators $IND1$ and $IND2$ defined in Section 10.6 are designed to guide the search for a suitable auxiliary vector $\mathbf{x}_k$, one that will be used to compute the definitive weights for the calibration estimator. Before the final decision with regard to $\mathbf{x}_k$ is taken, important guidance can be received by computing *both* indicators for a number of candidate auxiliary vectors. Both are functions of the auxiliary vector $\mathbf{x}_k$. If, for a certain $\mathbf{x}_k$, both indicators show higher values than for competing auxiliary vectors, then there is good reason to believe that this $\mathbf{x}_k$ produces the smallest bias for the calibration estimator, among the alternatives considered.

In the search for an $\mathbf{x}_k$-vector, $IND1$ and $IND2$ should be evaluated jointly. The two indicators complement each other. Together they provide a tool for comparing alternative $\mathbf{x}_k$-vectors within one and the same survey. What is of interest is the *change* in the values of $IND1$ and $IND2$ in moving from one $\mathbf{x}_k$-vector to another. When an x-variable is added, both are likely to increase. In the end, the bias of the estimator derived from the preferred $\mathbf{x}_k$ still remains an unknown quantity. It is an uncertainty that we must accept. A vector $\mathbf{x}_k$ that succeeds in satisfying Principle 1 would reduce the bias for all study variables. It is thus particularly important to monitor the progression of $IND1$ when the auxiliary vector is allowed to change.

Adding further variables (categorical or continuous) to $\mathbf{x}_k$ will in general increase the value of both $IND1$ and $IND2$. However, the expansion of $\mathbf{x}_k$ should be viewed with a critical eye. Associated with an expansion of $\mathbf{x}_k$ is the risk of some undesirable or abnormal values v_{sk}, which is to be avoided.

Minute increases in one or both of $IND1$ and $IND2$ must not be seen as an obligation to expand the $\mathbf{x}_k$-vector. In a regularly repeated survey it is desirable to keep the same $\mathbf{x}_k$ over a sequence of survey occasions. Good survey practice is not furthered by using an 'overextended' $\mathbf{x}_k$ or by changing it minimally from one survey occasion to the next.

This section uses two empirical studies to illustrate the gradual increments of $IND1$ and $IND2$. The first study is a Monte Carlo simulation, a continuation of the one reported in Section 7.7. The study uses the same population, KYBOK, and the same two response distributions as in Section 7.7. The average of the indicators $IND1$ and $IND2$ is computed over a large number of repeated outcomes s and r, for nine different estimators.

In the second study we return to the Survey on Life and Health described in Example 3.5. We compute the indicators $IND1$, $IND1$A and $IND2$ for a range of different auxiliary vectors $\mathbf{x}_k$. When this survey was designed at Statistics Sweden, the auxiliary vector was decided on by the informal analysis reported in Example 3.5, without the aid of the indicators. The purpose here is to compute the three indicators for a broader range of alternative auxiliary vectors.

A Monte Carlo simulation study

Both $IND1$ and $IND2$ are random quantities. For a fixed auxiliary vector $\mathbf{x}_k$, they depend on the chosen sample s and the response set r, both of which are random sets. Therefore it is appropriate to examine the behaviour of $IND1$ and $IND2$ 'in the long run', that is, on average over a long series of realizations of s and r. This is achieved by simulation, using different alternatives for the sampling design and for the response distribution.

For the simulation reported here, the population, KYBOK, and the two response distributions are the same as in Sections 7.7 and 8.3. We drew 10 000 SI samples s, each of size $n = 300$. For each sample we generated one response set r for each of the two response distributions described in Section 7.7.

For every realization of s and r, and for nine different estimators, we computed $IND1$ and $IND2$, as defined by (10.3) and (10.14), and then averaged the resulting 10 000 values to obtain the simulation averages $\overline{IND1}$ and $\overline{IND2}$, shown in Table 10.4. To compute $IND1$ and $IND2$ for a given estimator, we need the data $(\mathbf{x}_k, \mathbf{z}_k)$ for $k \in s$.

The simulation includes the eight estimators evaluated in Section 7.7: EXP, WC, PWA, RA, SEPRA, REG, SEPREG and TWOWAY. Table 10.1 identifies the pair $(\mathbf{x}_k, \mathbf{z}_k)$ corresponding to each of these estimators.

We also included the single-step calibration estimator SS:WC/REG defined by (8.7) and (8.8). This is based on the combined auxiliary vector $\mathbf{x}_k = \begin{pmatrix} x_k \\ \mathbf{x}_k^{\circ} \end{pmatrix}$ with $\mathbf{x}_k^{\circ} = \boldsymbol{\gamma}_k = (\gamma_{1k}, \ldots, \gamma_{pk}, \ldots, \gamma_{Pk})'$.

Table 10.4 Simulation mean of $IND1$ (multiplied by 10^3), simulation mean of $IND2$ (%) and simulation relative bias of $\hat{Y}_W$ (%) for two response distributions and different point estimators. Samples of size $n = 300$ drawn by SI. Nonresponse rate 14 %.

Response distribution	Estimator	$\overline{IND1}$ $\times 10^3$	$\overline{IND2}$ $\times 10^2$	$RB_{\mathrm{SIM}}(\hat{Y}_W)$ $\times 10^2$
Logit	Expansion (EXP)	0.0	0.0	5.0
	Weighting class (WC)	2.7	43.3	2.2
	Population weighting adjustment (PWA)	2.7	43.3	2.2
	Ratio (RA)	0.0	58.5	2.5
	Separate ratio (SEPRA)	2.7	82.5	0.7
	Regression (REG)	2.2	83.4	−0.6
	Separate regression (SEPREG)	6.0	88.1	−0.2
	Two-way classification (TWOWAY)	5.7	67.4	0.5
	Single-step (SS:WC/REG)	4.1	84.5	−0.6
Increasing exponential	Expansion (EXP)	0.0	0.0	9.4
	Weighting class (WC)	3.4	42.3	5.7
	Population weighting adjustment (PWA)	3.4	42.3	5.7
	Ratio (RA)	0.0	58.7	3.9
	Separate ratio (SEPRA)	2.3	81.0	2.0
	Regression (REG)	9.4	81.7	−2.7
	Separate regression (SEPREG)	18.3	88.1	−0.8
	Two-way classification (TWOWAY)	18.0	67.1	0.5
	Single-step (SS:WC/REG)	12.2	82.9	−2.4

An important objective of the exercise is to see how well $\overline{IND1}$ and $\overline{IND2}$ 'track' the relative bias of the nine estimators. To this end, Table 10.4 also shows the value of the simulation relative bias, $RB_{SIM}(\hat{Y}_W)$, taken from Tables 7.2 and 8.1.

If $IND1$ and $IND2$ function well as indicators, the simulation should confirm the following patterns:

- When the values of both $\overline{IND1}$ and $\overline{IND2}$ increase as a result of changing the estimator, we expect to see an accompanying decrease in the absolute value of $RB_{SIM}(\hat{Y}_W)$.
- When we rank the estimators by the value (from smallest to largest) of $\overline{IND1}$, the more important of the two indicators, we expect roughly the same rank order as when they are ranked by the absolute value (from largest to smallest) of $RB_{SIM}(\hat{Y}_W)$.
- Out of two or more estimators with roughly the same value on $\overline{IND1}$, we expect the one with the higher value of $\overline{IND2}$ to have the smaller absolute value of $RB_{SIM}(\hat{Y}_W)$.
- The order of preference (based on the value of $\overline{IND1}$ or on the absolute value of $RB_{SIM}(\hat{Y}_W)$) among the different estimators is expected to be roughly the same for both response distributions.
- For every given estimator, we expect $\overline{IND2}$ (but not necessarily $\overline{IND1}$) to give approximately the same value for both response distributions. The property is important, because in a real survey, the purpose of $IND2$ is to measure how well y is predicted by $\mathbf{x}$, irrespective of the unknown response distribution.

These desirable patterns are, by and large, confirmed by Table 10.4. We confirmed the 'tracking property' of $IND1$ for other response distributions, including some where θ_k depends on y_k.

The results in Table 10.4 prompt the following comments, which apply to both response distributions:

- From Section 10.6 we know that $IND1 = 0$ for both EXP and RA. In Table 10.4, RA has the higher value of $IND2$. We expect RA to have a smaller relative bias than EXP. This is confirmed by the simulation relative bias: $RB_{SIM}(\hat{Y}_W)$ is lower for RA.
- The estimators WC and PWA use the same auxiliary vector $\mathbf{x}_k$. Consequently, WC and PWA have a common value for $\overline{IND1}$ and for $\overline{IND2}$. These values are higher than for EXP, so we expect WC and PWA to have smaller bias than EXP. This is confirmed by the absolute values of the simulation relative bias, $RB_{SIM}(\hat{Y}_W)$.
- SEPREG and TWOWAY distinguish themselves by relatively high values of $\overline{IND1}$. We thus expect them to have less bias than the other seven estimators. This is confirmed by the absolute values of $RB_{SIM}(\hat{Y}_W)$.
- Five estimators, WC, PWA, SEPRA, REG and SS:WC/REG, have 'medium range' values of $\overline{IND1}$. Of those five, SEPRA, REG and SS:WC/REG show distinctly higher values of $\overline{IND2}$. Accordingly, we expect a smaller bias for

those three estimators, as compared to WC and PWA. This is confirmed by the absolute values of $RB_{\text{SIM}}(\hat{Y}_{\text{W}})$.

In summary, Table 10.4 supports the contention that inspection of $IND1$ and $IND2$ together can be helpful in identifying a preferred auxiliary vector $\mathbf{x}_k$. Thus $IND1$ and $IND2$ together provide a guide, although not a perfect guide, to the choice of $\mathbf{x}_k$-vector likely to generate the smallest bias. It should be emphasized that $IND1$ and $IND2$ are strictly tools for finding an appropriate $\mathbf{x}_k$-vector. Their mission is not to be flawless predictors of the value of the bias.

The interpretation of Table 10.4 is not entirely free of inconsistencies. For example, in comparing REG and SEPRA for the increasing exponential response distribution, we find that $IND1$ is markedly higher for REG, while $IND2$ is at about the same level for both. But the expected lower absolute value of $RB_{\text{SIM}}(\hat{Y}_{\text{W}})$ does not materialize for REG.

Because $IND1$ is computed as a variance (that is, as an average of squared deviations), it is sensitive to large values of v_{sk}. Outlying values of v_{sk} can occur, for example, if a response set contains one or more abnormally large elements. As argued elsewhere, it is part of the diagnostic effort to avoid $\mathbf{x}_k$-vectors that generate abnormally large or small weights, even though the number of such weights may be very small in the realized response set r.

Remark 10.6

In the simulation we also computed the indicator $IND1A$ defined by (10.8). (For $IND1A$, it is assumed that $\mathbf{x}_k$ is available for all $k \in s$.) It can be erratic when the auxiliary vector $\mathbf{x}_k$ contains a continuous variable. This is a reason for preferring $IND1$ to $IND1A$. In this simulation, the mean $\overline{IND1A}$ of $IND1A$ turned out to be negative, thus unacceptable, for RA and SEPRA. The explanation is that $\hat{\theta}_k$ as defined in (10.8) can exceed unity, perhaps considerably, for one or more elements in some samples s. This may happen when a continuous variable included in $\mathbf{x}_k$ has an unusually large value for one or a few elements, as compared to the mean of that variable for the other elements. To avoid this irregularity, an alternative is to compute $IND1A$ by (10.8) but with $\hat{\theta}_k$ determined instead by the logistic formula (8.16). □

The Survey on Life and Health

This study focuses on one single sample, namely, the sample taken for Statistics Sweden's Survey on Life and Health, as described in Example 3.5. We wish to illustrate the progression of the indicators $IND1$, $IND1A$ and $IND2$ as the auxiliary vector changes in the calibration estimator (6.11). In particular, we are interested in their increments as further auxiliary variables are added to the $\mathbf{x}_k$-vector.

Unlike the simulation study described earlier in this section, there are no repeated draws of samples and no measure of relative bias against which we can gauge the effectiveness of $IND1$, $IND1A$ and $IND2$.

We computed the values of the three indicators for a list of 21 potential auxiliary vectors x_k which do not exhaust all possibilities. All the variables used to form these vectors are categorical. The results are shown in Table 10.5. The indicator $IND2$ was computed for the particular study variable 'poor health'.

In Table 10.5, the '+' sign placed between two or more categorical variables indicates that they are 'strung out side-by-side' in the auxiliary vector x_k, in the manner of (7.19). The sign '×' indicates that the categorical variables are crossed in the auxiliary vector x_k.

The results in Table 10.5 prompt the following comments:

- The addition of a new categorical auxiliary variable increases the value of all three indicators $IND1$, $IND1A$ and $IND2$. This is illustrated particularly

Table 10.5   Values of $IND1$ (multiplied by 10^3), $IND1A$ (%) and $IND2$ (%) for different auxiliary vectors, based on data from Statistics Sweden's Survey on Life and Health.

Auxiliary vector	$IND1 \times 10^3$	$IND1A \times 10^2$	$IND2 \times 10^2$
Sex	16.7	1.4	0.1
Age group	42.6	3.5	1.0
Country of birth	11.1	0.7	0.1
Income class	11.7	0.9	0.4
Education level	4.4	0.4	0.4
Municipality	1.0	0.1	0.2
Civil status	26.8	2.2	0.0
Age group + Civil status	53.4	4.4	1.2
Age group × Civil status	54.2	4.4	1.2
Sex + Age group	60.6	4.7	1.1
Sex + Age group + Municipality	61.7	4.8	1.3
Sex + Age group + Municipality + Country of birth	70.5	5.3	1.4
Sex + Age group + Municipality + Country of birth + Education level	83.5	6.2	1.7
Sex + Age group + Municipality + Country of birth + Education level + Civil status	95.1	7.1	1.9
Sex + Age group + Municipality + Country of birth + Education level + Civil status + Income class	105.8	7.8	2.3
Sex × Age group	71.3	5.1	1.2
Sex × Age group × Municipality	83.2	5.9	2.2
Sex × Age group × Municipality + Country of birth	92.3	6.4	2.3
Municipality × Sex × Age group + Country of birth + Educational level	105.0	7.3	2.6
Sex × Age group × Municipality + Country of birth + Education level + Civil status	114.6	8.1	2.8
Sex × Age group × Municipality + Country of birth + Education level + Civil status + Income class	124.7	8.7	3.2

in the sequence of additions, one variable at a time, to *Sex + Age group* and to *Sex × Age group × Municipality*.

- Both *IND*1 and *IND*1A are constructed to help satisfy Principle 1. Here they agree closely in the ranking of the 21 alternative auxiliary vectors. Of the two, *IND*1 is considered more reliable. When $\mathbf{x}_k$ is made up entirely of categorical variables, as is the case here, *IND*1A is less prone to give erratic values than when $\mathbf{x}_k$ contains one or more continuous variables. The indicator *IND*2 was constructed with Principle 2 in mind. It, too, has an increasing tendency when the $\mathbf{x}_k$-vector expands.

- A comparison of the first seven rows of Table 10.5 shows that all three indicators designate *Age group* as the most powerful single variable. *Civil status* is also important for *IND*1 and *IND*1A. But *Age group + Civil status* is not the most powerful 'addition of two variables'; *Sex + Age group* gives higher values of *IND*1 and *IND*1A. There are some non-negligible interactions, for example, between sex and age group: $IND1 \times 10^3$ increases from 60.6 to 71.3 when the auxiliary vector changes from *Sex + Age group* into *Sex × Age group*. By contrast, the increases are negligible when *Age group + Civil status* is replaced by *Age group × Civil status*.

- The survey managers identified the most important domains to be those defined by the crossing *Sex × Age group × Municipality*. To respect Principle 3, the auxiliary vector should contain this three-factor crossing. The vector *Sex × Age group × Municipality* gives fairly high values of *IND*1 and *IND*1A. Further additions bring some improvement.

- When the survey was designed, the final auxiliary vector was fixed to be *Sex × Age group × Municipality + Country of birth + Educational level* (see Section 3.3). The choice was guided in part by a concern not to 'overextend' the auxiliary vector. If indicators had been used at the time, an analysis would have revealed that this final choice of auxiliary vector is not in line with a strategy of maximizing the indicator values. The further addition of income class and civil status gives higher values of the three indicators.

As remarked earlier, an unduly extensive $\mathbf{x}_k$-vector may compromise the quality of the resulting calibrated weights. Some weights may become abnormally high or low and upset the estimation. The variance estimation procedure explained in Chapter 11 may also be perturbed by abnormal weights. When several categorical auxiliary variables are available, we have recommended the principle of 'stringing out side-by side' as a means of avoiding the small cells and abnormal weights that may arise from a complete cross-classification.

However, in the survey on Life and Health, it was found that the cross-classification *Sex × Age group × Municipality* did not cause any extreme weights, neither did the most extensive $\mathbf{x}_k$-vector in Table 10.5, shown in the last row as *Sex × Age group × Municipality + Country of birth + Education level + Civil status + Income class*. For that $\mathbf{x}_k$-vector all of the weights v_{sk} given by (10.2) are contained in the interval [0.67, 3.01]. The largest weight is not abnormal by any standards.

A very similar distribution of weights is obtained when we string out all seven categorical variables side by side, so that the vector becomes *Sex + Age group +*

Municipality + *Country of birth* + *Education level* + *Civil status* + *Income class*. All weights v_{sk} are then contained in the interval [0.60, 2.64], which is quite normal. This vector has only slightly weaker values of the three indicators, as compared to *Sex* × *Age group* × *Municipality* + *Country of birth* + *Education level* + *Civil status* + *Income class*.

10.8. LITERATURE REVIEW

There exists a large literature on the selection of 'the best' auxiliary information. This selection is discussed in the literature from different angles. Some of the recommendations in these papers are relevant also for the topic of this book. We present a brief literature review. The objective is only give some examples of the issues brought up by different authors. It is important to note that the material quoted comes from publications dated prior to the new work reported in this book.

The many opinions that have been expressed cannot be easily summarized in a few firm conclusions or recommendations. Instead we concentrate on five important themes in the selection of auxiliary variables, and review some of the literature under these headings. The review is not in any way exhaustive.

Reduction of the variance

Nascimento Silva and Skinner (1997) focus on a reduction of sampling error, in the ideal situation of no nonresponse. Their intention is to select the 'optimal' set of auxiliary variables from a rather large set of potentially useful variables. They compare different ways to carry out this selection and finally recommend an adaptive technique. It is based on the strength of relationship found in the sample between the study variable and the auxiliary variables. In other words, the selected set of auxiliary variables is sample-dependent.

By contrast, in their study of auxiliary variables for use in estimation for the Canadian census, Bankier *et al.* (1992) consider a selection based strictly on the interrelationships among the auxiliary variables themselves.

The selection of auxiliary information is also important for the construction of an efficient sampling design. Although this question falls outside the main scope of this book, we note that several authors have addressed the selection problem from this angle. For example, Kish and Anderson (1978) discuss alternative ways to stratify the population for a survey with many study variables and with many purposes. They stress the importance of using a rather large number of stratifiers with relatively few categories for each stratifier as opposed to many categories for each of a small number of stratifiers:

> ... *the advantages of several stratifiers are much greater for multipurpose surveys* ... *For any stratifier, the gains in reducing the variance within strata show rapidly diminishing returns with few strata.*

A similar principle is likely to work well for the construction of a set of poststrata, when the purpose is to reduce the sampling variance.

Reduction of the bias

We have noted that the bias of the calibration estimator $\hat{Y}_W$ may be considerably reduced when powerful auxiliary variables are available. In particular, Principle 1 in Section 10.2 states that the response influence should be well explained by the instrument vector or the auxiliary vector.

There is a vast literature, based to a large extent on empirical evidence, on patterns of correlation that may exist between response propensity and potential auxiliary variables. Studies of surveys on individuals have shown that lower response rates are typically to be expected for the following categories of respondents: metropolitan residents; single people; members of childless households; older people; divorced/widowed people; persons with lower educational attainment; and self-employed people (see Holt and Elliot, 1991; Lindström, 1983). Issues in response patterns for business surveys are discussed, for example, in Willimack *et al.* (2002).

Estimation for domains

Principle 3 in Section 10.2 stresses the importance of auxiliary information about groups that overlap considerably with the domains of interest, particularly the most important domains. Such information is not always at hand in the survey.

The simultaneous estimation for several sets of domains creates some special problems. Assume that estimates are needed for two sets of domains, defined, respectively, by the groups $p = 1, \ldots, P$ and the groups $h = 1, \ldots, H$, such as those referred to in the discussion of the two-way classification in Section 7.6. The number of responding elements may be very small in many of the $P \times H$ cells arising from a complete crossing of the two groupings. We could then base the calibration on different one-way classification vectors, $\mathbf{x}_k = (\gamma_{1k}, \ldots, \gamma_{pk}, \ldots, \gamma_{Pk})'$ for the first set of domains and $\mathbf{x}_k = (\delta_{1k}, \ldots, \delta_{hk}, \ldots, \delta_{Hk})'$ for the second set of domains. But one may consider it a disadvantage that it gives two different weighting systems, one for the first set of domains, and another for the second set. A remedy is to use instead a compromise vector of the two-way classification type, $\mathbf{x}_k = (\gamma_{1k}, \ldots, \gamma_{pk}, \ldots, \gamma_{Pk}, \delta_{1k}, \ldots, \delta_{hk}, \ldots, \delta_{H-1,k})'$ and to compute a single weighting system with this vector. This solution may be somewhat short of ideal for either of the two sets of domains; however, as Lundström (1996) illustrates for a survey at Statistics Sweden, the variance of the domain estimates tends to be only slightly larger for the two-way classification vector, as compared to the alternative with two separate vectors.

Interaction between the objective of variance reduction and the objective of bias reduction

Comparatively few papers discuss the choice of auxiliary information for the dual purpose of variance reduction and bias reduction. However, many statisticians are aware that the choice of auxiliary information should serve this dual purpose.

Little (1986) discusses the choice of suitable weighting classes when a large amount of information is available. In the *predicted mean stratification* method, the study variable y is regressed on $\mathbf{x}$ in the response set r and then the weighting classes are constructed by grouping the predicted y-values for the sample. In the *response propensity stratification* method, the weighting classes are based on intervals of the estimated response probabilities, called response propensity scores and obtained by logistic or probit regression fitting. Little (1986) notes that:

... *predicted mean stratification has the virtue of controlling both the bias and variance...; response propensity stratification controls ... bias, but yields estimates ... that may have large variance. The latter is particularly true when the response propensity is largely determined by variables that are not associated with y.*

On the question of the dual purpose, Oh and Scheuren (1983) and Särndal and Swensson (1987) express somewhat different views. Oh and Scheuren (1983) state:

A seemingly robust approach is to choose the subgroups such that for the variable(s) to be analysed, the within-group variation for nonrespondents is small (and the between-group mean differences are large); then, even if the response mechanism is postulated incorrectly, the bias impact will be small.

Särndal and Swensson (1987) react to this as follows:

In our opinion, one must separate the role of the RHGs [response homogeneity groups] from that of other information ... recorded for $k \in s$. Two different concepts are involved. The sole criterion for the RHGs should be that they eliminate bias as far as possible. Every effort should be made, and a prior knowledge used, to settle on groups likely to display response homogeneity. But in addition it is imperative to measure, for $k \in s$, a concomitant vector $\mathbf{x}_k$, that will reduce variance and give added protection against bias. Groups that eliminate or reduce bias are not necessarily variance reducing, and, contrary to what the quotation seems to suggest, the criterion of maximizing between-to-within variation in y does not necessarily create groups that work well for removing bias.

Särndal and Swensson (1987) argue in principle for the use of two different 'concepts', but they admit that practical problems can arise:

... *in order to eliminate bias due to nonresponse, it is vital to identify the true response model; as this is usually impossible, bias can be greatly reduced if powerful explanatory x-variables can be found and incorporated in a regression-type estimator.*

In a similar vein, Bethlehem (1988) states:

... *it is very important to look for good stratification variables that will reduce both variance and bias.*

Some authors warn that the simultaneous reduction of the variance and the bias may represent conflicting ambitions. Kalton and Maligalig (1991) express this in the following way:

In general, a price paid for adjustment cell weighting is a loss of precision in the survey estimators. There is a trade-off to be made between bias reduction and an increase in variance arising from the variation in the weights. The increase is not great when the variation in weights is modest, but it rises rapidly as the variation increases.... Common techniques for restricting the variation in weights are to collapse cells and to trim the weights ... Since cells with small sample sizes often give rise to large variation in weights, minimum sample sizes in the cells are often specified (e.g., 25, 30 or 50).

Problems generated by the random nature of the sample

Another question discussed in the literature is whether the current sample information should be allowed to direct the choice of auxiliary information. It is well known that sample-based selection of auxiliary variables may affect important properties of the point estimator, particularly its variance. For example, Bethlehem (1988) points to the desirability of basing the selection of auxiliary variables on information other than what is observed in the current sample:

The choice of stratification variables cannot be made solely on the basis of the available observations. Over- or under-representation of some groups can mislead us about the relationship between the target and the stratification variable. There has to be additional information about the homogeneity of the target variable.

The situation is favourable in the case of a regularly repeated survey, where historical data exist, in addition to the information from current sample. However, for first-time surveys, there is no historical precedent. One must act exclusively on the data from the current occasion.

Nascimento Silva and Skinner (1997) study the selection of auxiliary variables for the case of complete response. They find that estimators based on 'best possible selection' of auxiliary variables *for each chosen sample* can be effective. Such sample-dependent selection of auxiliary variables can give a lower MSE for a GREG estimator, as compared to an a priori fixed set of auxiliary variables. The selection is thus contingent on an inspection of the chosen sample.

A simple and commonly used sample-based technique is the collapsing of groups, with a collapsing rule based on the number of respondents in the groups of the sample. Small groups should, generally speaking, be avoided. Procedures for variance estimation may degenerate when the number of respondents in a group is extremely small.

Kalton and Kasprzyk (1986) discuss the undesirable effects that an excessive variability in the weights can have on the variance. They consider the procedure in which the final weights result from two adjustments. Two subweights are used. The first 'compensates for unequal response rates in different sample weighting classes', and the second 'makes the weighted sample distribution for certain characteristics ... conform to the known population distribution for those characteristics'. This can be described as a reweighting step followed by a poststratification step, conforming essentially to what we have called procedure two-step B. They recommend inspecting the final weights and, if some are too large, collapsing groups or 'trimming the weights' in order to avoid unacceptably large variance.

Inspection of the distribution of the final weights is an important step that should always be undertaken. This holds also for the calibration approach to weighting, as discussed in this book. That is, the variation of weights such as g_k, v_k and v_{sk} should be analysed. When extreme weights occur, the first question to examine is whether two or more auxiliary variables measure essentially the same concept. Such collinearity may cause unstable weights. Variables causing collinearity should be excluded from the auxiliary vector.

Methods exist for restricting weights. One approach is to make sure that they lie within a prespecified interval. Procedures of this kind are available, for example, in CALMAR and CLAN97.

CHAPTER 11

Variance and Variance Estimation

11.1. VARIANCE ESTIMATION FOR THE CALIBRATION ESTIMATOR

This fairly technical chapter presents the procedure for estimating the variance of a calibration estimator $\hat{Y}_W$. The variance estimate is needed for two purposes: to indicate the precision of $\hat{Y}_W$; and to compute a confidence interval centred on $\hat{Y}_W$. The data available for this purpose are the y-data observed for the responding elements and any auxiliary information that may be helpful.

Proposition 11.1 in this section presents the variance estimator. The rationale for the technique is explained later in the chapter. The computational aspect and the use of software for variance estimation are discussed in Section 11.6. Proposition 11.1 is formulated for the general case of InfoUS, where auxiliary information exists at both the sample level and the population level. As in Section 6.2, the information has the following features:

(i) $\mathbf{x}_k^*$ is an observed or otherwise known vector value for every $k \in r$ and the total $\mathbf{X}^* = \sum_U \mathbf{x}_k^*$ is known (or computable, as when $\mathbf{x}_k^*$ is known for every $k \in U$).

(ii) $\mathbf{x}_k^\circ$ is an observed or otherwise known vector value for every $k \in r$ and $\hat{\mathbf{X}}^\circ = \sum_s d_k \mathbf{x}_k^\circ$ is computable, as it is when $\mathbf{x}_k^\circ$ is known for every $k \in s$.

The auxiliary vector for InfoUS is $\mathbf{x}_k = \begin{pmatrix} \mathbf{x}_k^* \\ \mathbf{x}_k^\circ \end{pmatrix}$, and we let $\mathbf{z}_k$ denote an instrument vector of matching dimension. In a majority of the cases we consider, the standard specification $\mathbf{z}_k = \mathbf{x}_k$ applies. Variance estimation for InfoU and InfoS is obtained as special cases of Proposition 11.1 by letting $\mathbf{x}_k = \mathbf{x}_k^*$ and $\mathbf{x}_k = \mathbf{x}_k^\circ$, respectively. We concentrate in this section on the single-step procedure. The two-step A and two-step B procedures are considered in Section 11.4.

In single-step, we calibrate directly on the information $\mathbf{X} = \begin{pmatrix} \mathbf{X}^* \\ \hat{\mathbf{X}}^\circ \end{pmatrix}$ and obtain the calibrated weights given by (6.14) with $d_{\alpha k} = d_k$, namely,

$$w_k = d_k v_k, \qquad v_k = 1 + \boldsymbol{\lambda}_r' \mathbf{z}_k \qquad (11.1)$$

Estimation in Surveys with Nonresponse   C.-E. Särndal and S. Lundström
© 2005 John Wiley & Sons, Ltd

where $\lambda'_r = \left(\mathbf{X} - \sum_r d_k \mathbf{x}_k \right)' \left(\sum_r d_k \mathbf{z}_k \mathbf{x}'_k \right)^{-1}$. The resulting single-step calibration estimator is

$$\hat{Y}_{\mathrm{W}} = \sum_r w_k y_k. \tag{11.2}$$

The following proposition presents the variance estimator for $\hat{Y}_{\mathrm{W}}$ as the sum of an estimated sampling variance component and an estimated nonresponse variance component. The quantities $d_k = 1/\pi_k$ and $d_{k\ell} = 1/\pi_{k\ell}$ are needed for the computation, where π_k and $\pi_{k\ell}$ are the inclusion probabilities (of first and second order, respectively) generated by the sampling design $p(s)$. The $d_{k\ell}$ will intervene in the variance computation; to avoid the double sum, it is possible to resort to an approximation that dispenses with the $d_{k\ell}$ in the manner examined in Section 4.4.

Proposition 11.1

The variance of $\hat{Y}_{\mathrm{W}}$ is estimated by

$$\hat{V}(\hat{Y}_{\mathrm{W}}) = \hat{V}_{\mathrm{SAM}} + \hat{V}_{\mathrm{NR}}. \tag{11.3}$$

The estimated sampling variance component is

$$\hat{V}_{\mathrm{SAM}} = \sum \sum_r (d_k d_\ell - d_{k\ell})(v_k \hat{e}^*_k)(v_\ell \hat{e}^*_\ell) - \sum_r d_k (d_k - 1) v_k (v_k - 1)(\hat{e}^*_k)^2 \tag{11.4}$$

and the estimated nonresponse variance component is

$$\hat{V}_{\mathrm{NR}} = \sum_r v_k (v_k - 1)(d_k \hat{e}_k)^2 \tag{11.5}$$

with

$$\hat{e}^*_k = y_k - (\mathbf{x}^*_k)' \mathbf{B}^*_{r;dv},$$

$$\hat{e}_k = y_k - \mathbf{x}'_k \mathbf{B}_{r;dv} = y_k - (\mathbf{x}^*_k)' \mathbf{B}^*_{r;dv} - (\mathbf{x}^\circ_k)' \mathbf{B}^\circ_{r;dv}$$

in which

$$\mathbf{B}_{r;dv} = \begin{pmatrix} \mathbf{B}^*_{r;dv} \\ \mathbf{B}^\circ_{r;dv} \end{pmatrix} = \left(\sum_r d_k v_k \mathbf{z}_k \mathbf{x}'_k \right)^{-1} \left(\sum_r d_k v_k \mathbf{z}_k y_k \right). \tag{11.6}$$

□

The justification for the variance estimator in Proposition 11.1 is given later in this chapter.

Proposition 11.1 can be applied to the special cases InfoU and InfoS. For InfoU, the auxiliary vector is $\mathbf{x}_k = \mathbf{x}^*_k$, and the residuals in Proposition 11.1 become

$$\hat{e}^*_k = \hat{e}_k = y_k - (\mathbf{x}^*_k)' \left(\sum_r d_k v_k \mathbf{z}^*_k (\mathbf{x}^*_k)' \right)^{-1} \left(\sum_r d_k v_k \mathbf{z}^*_k y_k \right)$$

where $\mathbf{z}^*_k$ is a vector with dimension matching that of $\mathbf{x}^*_k$. For InfoS, the auxiliary vector is $\mathbf{x}_k = \mathbf{x}^\circ_k$, and

$$\hat{e}^*_k = y_k, \qquad \hat{e}_k = y_k - (\mathbf{x}^\circ_k)' \left(\sum_r d_k v_k \mathbf{z}^\circ_k (\mathbf{x}^\circ_k)' \right)^{-1} \left(\sum_r d_k v_k \mathbf{z}^\circ_k y_k \right)$$

where $\mathbf{z}^\circ_k$ and $\mathbf{x}^\circ_k$ have matching dimensions.

We can use the variance estimator (11.3) to compute a confidence interval for Y. Aiming at an approximately $1 - \alpha$ confidence level, we compute $\hat{Y}_W \pm z_{\alpha/2}\sqrt{\hat{V}(\hat{Y}_W)}$, where $z_{\alpha/2}$ is the standard normal score. For example, $z_{\alpha/2} = 1.96$ for an intended confidence level of approximately $1 - \alpha = 95\,\%$. The fact that $\hat{Y}_W$ is approximately (not exactly) normally distributed is one reason why the confidence level $1 - \alpha$ is not attained exactly, but a more troublesome reason is that $\hat{Y}_W$ may have a nonnegligible bias due to nonresponse. To trust the interval, one must be reasonably assured that the bias of $\hat{Y}_W$ is near zero. If it is not, the interval tends to be off-centre. The resulting damage to the coverage rate is illustrated empirically in Section 11.5. The coverage rate achieved may fall considerably short of the intended confidence level $1 - \alpha$, and the interval becomes invalid. An efficient bias reduction, using the methods in Chapters 9 and 10, is therefore a necessary prerequisite for a meaningful confidence interval.

Despite some bias in the point estimator, the confidence interval method still has merit as long as the *bias ratio* is limited. The bias ratio of an estimator is defined as the bias divided by the standard error. On the assumption of a normally distributed estimator and a bias ratio equal to 0.5, the true confidence level (the coverage probability) of a nominal 95 % interval (thus based on the normal score 1.96) drops, but only to 92 %. It falls only 3 % short of the nominal 95 % level. This is discussed, for example, in Cochran (1977, Section 1.8) and in Särndal *et al.* (1992, Section 5.2).

The general formulae in Proposition 11.1 become more readily interpretable if we apply them to specific and familiar sampling designs. Consider the important special case of STSI, such that stratum h is sampled (with SI) at the rate n_h/N_h. The sampling weights are $d_k = N_h/n_h$ for all k in stratum h. From (11.1) we have

$$v_k = 1 + \left(\mathbf{X} - \sum_{h=1}^{H} \frac{N_h}{n_h} \sum_{r_h} \mathbf{x}_k \right)' \left(\sum_{h=1}^{H} \frac{N_h}{n_h} \sum_{r_h} \mathbf{z}_k \mathbf{x}_k' \right)^{-1} \mathbf{z}_k \qquad (11.7)$$

for $k \in r_h$. The estimated variance components (11.4) and (11.5) become

$$\hat{V}_{\mathrm{SAM}} = \sum_{h=1}^{H} N_h^2 \left(1 - \frac{n_h}{N_h} \right) \frac{1}{n_h} \frac{1}{n_h - 1} \left[\sum_{r_h} (v_k \hat{e}_k^*)^2 - \frac{1}{n_h} \left(\sum_{r_h} v_k \hat{e}_k^* \right)^2 \right]$$

$$- \sum_{h=1}^{H} \frac{N_h}{n_h} \left(\frac{N_h}{n_h} - 1 \right) \sum_{r_h} v_k (v_k - 1)(\hat{e}_k^*)^2 \qquad (11.8)$$

and

$$\hat{V}_{\mathrm{NR}} = \sum_{h=1}^{H} \left(\frac{N_h}{n_h} \right)^2 \sum_{r_h} v_k (v_k - 1) \hat{e}_k^2 \qquad (11.9)$$

where v_k is given by (11.7) and the residuals $\hat{e}_k^*$ and $\hat{e}_k$ are as specified in Proposition 11.1 with

$$\mathbf{B}_{r;dv} = \begin{pmatrix} \mathbf{B}_{r;dv}^* \\ \mathbf{B}_{r;dv}^\circ \end{pmatrix} = \left(\sum_{h=1}^{H} \frac{N_h}{n_h} \sum_{r_h} v_k \mathbf{z}_k \mathbf{x}_k' \right)^{-1} \sum_{h=1}^{H} \frac{N_h}{n_h} \sum_{r_h} v_k \mathbf{z}_k y_k.$$

Expressions still more in line with 'intuitive expectations' materialize if we examine InfoU with $\mathbf{x}_k = \mathbf{x}_k^* = \boldsymbol{\gamma}_k = \mathbf{z}_k$, where $\boldsymbol{\gamma}_k$ is the vector of dimension H that indicates classes which coincide with the H strata. Then (11.1) results in a straight expansion of the design weights $d_k = N_h/n_h$ by the inverse of the stratum response rate, so that $v_k = n_h/m_h$ for all $k \in r_h$, where m_h is the number of responding elements out of n_h sampled in stratum h. We insert $v_k = n_h/m_h$ into (11.8) and (11.9) and note that the appropriate residuals are $\hat{e}_k^* = \hat{e}_k = y_k - \bar{y}_{r_h}$ for $k \in r_h$. After some algebra, and use of the close approximations $n_h/(n_h - 1) \approx 1$ and $m_h/(m_h - 1) \approx 1$, we obtain from (11.8) and (11.9),

$$\hat{V}_{\text{SAM}} \approx \sum_{h=1}^{H} N_h^2 \left(\frac{1}{n_h} - \frac{1}{N_h} \right) S_{yr_h}^2$$

$$\hat{V}_{\text{NR}} \approx \sum_{h=1}^{H} N_h^2 \left(\frac{1}{m_h} - \frac{1}{n_h} \right) S_{yr_h}^2,$$

$$\hat{V}(\hat{Y}_{\text{W}}) = \hat{V}_{\text{SAM}} + \hat{V}_{\text{NR}} \approx \sum_{h=1}^{H} N_h^2 \left(\frac{1}{m_h} - \frac{1}{N_h} \right) S_{yr_h}^2$$

with $S_{yr_h}^2 = \sum_{r_h} (y_k - \bar{y}_{r_h})^2/(m_h - 1)$. These results are easy to interpret. The expression for $\hat{V}_{\text{SAM}}$ confirms the simple random selection of n_h elements from N_h in stratum h. The expression for $\hat{V}_{\text{NR}}$ embodies the idea of a conditional simple random selection of m_h from the n_h sampled elements in stratum h. In other words, the second variance component is computed *as if* the respondents are a simple random selection within each stratum. In practice we cannot know how well this approximates the truth, but it happens to be the interpretation that, for lack of more extensive information, accompanies the use of $\mathbf{x}_k = \mathbf{x}_k^* = \boldsymbol{\gamma}_k$ as auxiliary vector.

Several remarks are appropriate before examining the background and the uses of Proposition 11.1.

Remark 11.1

As argued earlier, a certain restrictiveness is recommended in allowing x-variables into the $\mathbf{x}_k$-vector. A general policy, mentioned in Remark 6.1, is to avoid x-variables that could cause undesirable weights. Near collinearity among x-variables can be a cause of abnormal weights. If backward elimination is practised to produce the final auxiliary vector, one should rid the $\mathbf{x}_k$-vector of those x-variables that are essentially redundant and may cause undesirable weights. Inspection of the set of weights v_k for $k \in r$ is always recommended. A few large weights may occur when the $\mathbf{x}_k$-vector contains one or more continuous x-variables with skewed distributions. A prudent approach is then to form size groups out of such x-variables.

In particular, undesirable weights v_k may unduly influence the variance estimator in Proposition 11.1. The computation of $\mathbf{B}_{r;dv}$ in (11.6) requires the inverse of the matrix $\sum_r d_k v_k \mathbf{z}_k \mathbf{x}_k'$. If $d_k v_k$ turns out to be negative for one or a few influential elements k, the matrix could become singular or nearly so, with erratic residuals $\hat{e}_k^*$ and $\hat{e}_k$ as a result. The problem is avoided by a restrictive attitude in allowing x-variables into the auxiliary vector. $\qquad\square$

Remark 11.2

We noted in Remark 4.1 that the estimated variance (4.18) of the GREG estimator has some tendency to underestimation in not-so-large samples. A similar weakness affects (11.3). The underestimation is attenuated by computing (11.5) with modified residuals that account for a loss of degrees of freedom, similarly to the way explained in Remark 4.1. $\qquad\square$

Remark 11.3

Proposition 11.1 also applies to estimation for domains. For the auxiliary information (i) and (ii) at the beginning of this section, we estimate the domain total $Y_q = \sum_{U_q} y_k$ by $\hat{Y}_{q\mathrm{W}} = \sum_U w_k y_{qk} = \sum_{U_q} w_k y_k$, where y_{qk} defined by (4.4) is the value of the domain-specific y-variable. The variance estimator for $\hat{Y}_{q\mathrm{W}}$ is given by (11.3)–(11.6) if we replace y_k by y_{qk} throughout. That is, $\mathbf{B}_{r;dv}$ is replaced by

$$\mathbf{B}_{(q)r;dv} = \begin{pmatrix} \mathbf{B}^*_{(q)r;dv} \\ \mathbf{B}^\circ_{(q)r;dv} \end{pmatrix} = \left(\sum_r d_k v_k \mathbf{z}_k \mathbf{x}_k' \right)^{-1} \left(\sum_r d_k v_k \mathbf{z}_k y_{qk} \right).$$

The residuals $\hat{e}_k^*$ and $\hat{e}_k$ change consequentially; they become domain-dependent. $\qquad\square$

11.2. AN ESTIMATOR FOR IDEAL CONDITIONS

The rationale for the variance estimator in Proposition 11.1 comes from a parallel that we can draw with estimation under the more favourable but impossible condition of a known response distribution. In that case, the selection occurs in two fully controlled phases. The first phase produces the sample s as before, with the known sampling weights $d_k = 1/\pi_k$. The second phase selects a response set r from s, where the response distribution $q(r|s)$ has the known first- and second-order response probabilities

$$\Pr(k \in r|s) = \theta_k, \qquad \Pr(k \ \& \ \ell \in r|s) = \theta_{k\ell}.$$

Let us pretend in this section that $q(r|s)$ is known, and that elements respond independently, so that $\theta_{k\ell} = \theta_k \theta_\ell$ for all $k \neq \ell$ and $\theta_{k\ell} = \theta_{kk} = \theta_k$ for all $k = \ell$. The response influence $\phi_k = 1/\theta_k$ is known under the conditions in this section. With the same information as in $\hat{Y}_{\mathrm{W}}$ given by (11.2), $\mathbf{X} = \begin{pmatrix} \mathbf{X}^* \\ \hat{\mathbf{X}}^\circ \end{pmatrix}$, we construct a two-phase generalized regression estimator of $Y = \sum_U y_k$, to be denoted $\hat{Y}_{\mathrm{G2p}}$. (To be more exact, $\hat{Y}_{\mathrm{G2p}}$ qualifies an estimator only in the ideal world where the response distribution is known. We are interested in $\hat{Y}_{\mathrm{G2p}}$ only as a tool in constructing a variance estimator for $\hat{Y}_{\mathrm{W}}$.)

With the information defined by (i) and (ii) in Section 11.1, we create the nearly unbiased estimator

$$\hat{Y}_{\text{G2p}} = \sum_r w_{\text{G2p}k} y_k \tag{11.10}$$

with the calibrated weights

$$w_{\text{G2p}k} = d_k \phi_k g_{k\phi}, \qquad g_{k\phi} = 1 + \boldsymbol{\lambda}'_r \mathbf{z}_k \tag{11.11}$$

where $\boldsymbol{\lambda}'_r = \left(\mathbf{X} - \sum_r d_k \phi_k \mathbf{x}_k \right)' \left(\sum_r d_k \phi_k \mathbf{z}_k \mathbf{x}'_k \right)^{-1}$.

The two phases of selection give element k the combined selection weight $(1/\pi_k) \times (1/\theta_k) = d_k \phi_k$, and $g_{k\phi}$ is a calibration adjustment. In this idealized world, we are rid of the problem of unknown response probabilities. Following linearization, it is easy to derive the two components of the approximate variance of $\hat{Y}_{\text{G2p}}$. This leads directly to an estimated variance, $\hat{V}(\hat{Y}_{\text{G2p}})$, which depends on the ϕ_k. The derivation of $\hat{V}(\hat{Y}_{\text{G2p}})$, deferred to the end-of-chapter appendix, entails isolating the principal term of the variance, followed by an estimation of that term. Here we only cite the expression for $\hat{V}(\hat{Y}_{\text{G2p}})$, because it is the blueprint for the computable variance estimator in Proposition 11.1. The estimated variance of $\hat{Y}_{\text{G2p}}$ is

$$\hat{V}(\hat{Y}_{\text{G2p}}) = \hat{V}_{\text{SAM}} + \hat{V}_{\text{NR}} \tag{11.12}$$

where

$$\hat{V}_{\text{SAM}} = \sum \sum_r (d_k d_\ell - d_{k\ell})(\phi_k \hat{e}^*_{k\phi})(\phi_\ell \hat{e}^*_{\ell\phi})$$
$$- \sum_r d_k(d_k - 1)\phi_k(\phi_k - 1)(\hat{e}^*_{k\phi})^2, \tag{11.13}$$

$$\hat{V}_{\text{NR}} = \sum_r \phi_k(\phi_k - 1)(d_k \hat{e}_{k\phi})^2. \tag{11.14}$$

The residuals are given by

$$\hat{e}^*_{k\phi} = y_k - (\mathbf{x}^*_k)' \mathbf{B}^*_{r;d\phi}, \qquad \hat{e}_{k\phi} = y_k - \mathbf{x}'_k \mathbf{B}_{r;d\phi} = y_k - (\mathbf{x}^*_k)' \mathbf{B}^*_{r;d\phi} - (\mathbf{x}^\circ_k)' \mathbf{B}^\circ_{r;d\phi}$$

where

$$\mathbf{B}_{r;d\phi} = \begin{pmatrix} \mathbf{B}^*_{r;d\phi} \\ \mathbf{B}^\circ_{r;d\phi} \end{pmatrix} = \left(\sum_r d_k \phi_k \mathbf{z}_k \mathbf{x}'_k \right)^{-1} \left(\sum_r d_k \phi_k \mathbf{z}_k y_k \right). \tag{11.15}$$

When ϕ_k is replaced by $\hat{\phi}_k = v_k$ in (11.12)–(11.15) we obtain the variance estimator for $\hat{Y}_{\text{W}}$ given in Proposition 11.1. The justification for the replacement is made explicit in the following section.

11.3. A USEFUL RELATIONSHIP

There is an obvious kinship between $\hat{Y}_{\text{W}}$ given by (11.2) and $\hat{Y}_{\text{G2p}}$ given by (11.10): both weight systems are calibrated to the same input of information, $\mathbf{X} = \begin{pmatrix} \mathbf{X}^* \\ \hat{\mathbf{X}}^\circ \end{pmatrix}$. This information is the principal determinant of the variance of both estimators. A difference is that the weights of $\hat{Y}_{\text{G2p}}$ depend explicitly on the influences $\phi_k = 1/\theta_k$; those of $\hat{Y}_{\text{W}}$ do not.

We would be in a position to use $\hat{Y}_{G2p}$ and its variance estimator $\hat{V}(\hat{Y}_{G2p})$ if the ϕ_k were known. Failing this, we cannot compute one or the other. On the other hand, $\hat{Y}_W$ given by (11.2) is readily computable, but an attempt to assess its variance runs into the obstacle of an unknown response distribution.

We justify the variance estimator for $\hat{Y}_W$ in Proposition 11.1 as follows. The idealized estimator $\hat{Y}_{G2p}$ given by (11.10) and (11.11) depends on the unknown ϕ_k, and so does the corresponding variance estimator (11.12). We create proxy values $\hat{\phi}_k$ for ϕ_k such that when $\hat{\phi}_k$ is substituted for ϕ_k in $\hat{Y}_{G2p}$, then $\hat{Y}_{G2p}$ becomes equal to $\hat{Y}_W$. (These $\hat{\phi}_k$ will be slightly different from the $\hat{\phi}_k$ given by (10.2), for reasons that will become apparent.) We express this equality as $\hat{Y}_{G2p|\phi=\hat{\phi}} = \hat{Y}_W$. It holds for the particular s and r on which the proxies $\hat{\phi}_k$ depend. It is reasonable to make the same substitution in the variance estimator (11.12). The result is the variance estimator for $\hat{Y}_W$ given by (11.3) in Proposition 11.1. We first establish the equality of $\hat{Y}_W$ and $\hat{Y}_{G2p}$.

Proposition 11.2

The equality $\hat{Y}_{G2p|\phi=\hat{\phi}} = \hat{Y}_W$ holds for $\phi_k = \hat{\phi}_k = v_k$, where v_k is given by (11.1).

□

Proof. Denote by $g_{k\hat{\phi}}$ the value of $g_{k\phi}$ in (11.11) when we set $\phi_k = \hat{\phi}_k = v_k$ given by (11.1). Then, using the calibration property $\sum_r d_k v_k \mathbf{x}_k = \mathbf{X}$, we find

$$g_{k\hat{\phi}} = 1 + \left(\mathbf{X} - \sum_r d_k v_k \mathbf{x}_k \right)' \left(\sum_r d_k v_k \mathbf{z}_k \mathbf{x}_k' \right)^{-1} \mathbf{z}_k = 1$$

for all k. Therefore, for any fixed s and r,

$$\hat{Y}_{G2p|\phi=\hat{\phi}} = \sum_r d_k v_k g_{k\hat{\phi}} y_k = \sum_r d_k v_k y_k = \sum_r w_k y_k = \hat{Y}_W.$$

□

Substituting $\hat{\phi}_k = v_k$ for ϕ_k in the variance estimator (11.12) for $\hat{Y}_{G2p}$, we obtain the variance estimator (11.3) for $\hat{Y}_W$. The outcome of the technique is that neither the point estimator $\hat{Y}_W$ nor its variance estimator requires the unknown influences ϕ_k.

What can be claimed for this technique? We know that $\hat{Y}_{G2p}$ has nearly zero bias, obtained as an average over all samples s and all response sets r, when the response distribution and its influences ϕ_k are known. We have established an equality between $\hat{Y}_{G2p}$ and $\hat{Y}_W$ *only* for a particular (s, r). It does not follow that $\hat{Y}_W$ is unbiased over all samples s and all response sets r. In fact, we know by Proposition 9.1 that the bias of $\hat{Y}_W$ is $B_{pq}(\hat{Y}_W) \approx -\sum_U (1 - \theta_k) e_{\theta k}$, usually a nonzero quantity.

Although the proxies $\hat{\phi}_k$ are less than perfect substitutes for the unknown influences ϕ_k, the variance estimate (11.3), obtained by the substitution procedure, will give a valuable indication of the precision of $\hat{Y}_W$. It can be used for confidence interval computation. If the bias of $\hat{Y}_W$ is modest, the confidence interval will be approximately valid, with a true confidence level (coverage rate) roughly equal to the intended $1 - \alpha$.

Remark 11.4

The proxies $\hat{\phi}_k = v_k$ used here differ slightly from the proxies $\hat{\phi}_k = v_{sk}$ used in Chapter 10. The proxies $\hat{\phi}_k = v_k$ given by (11.1) have the calibration property

$$\sum_r d_k \hat{\phi}_k \mathbf{x}_k = \sum_r d_k v_k \mathbf{x}_k = \begin{pmatrix} \mathbf{X}^* \\ \hat{\mathbf{X}}^\circ \end{pmatrix} = \mathbf{X}$$

where $\mathbf{X}^* = \sum_U \mathbf{x}_k^*$ and $\hat{\mathbf{X}}^\circ = \sum_s d_k \mathbf{x}_k^\circ$. The alternative proxies $\hat{\phi}_k = v_{sk}$ given by (10.2) have the calibration property

$$\sum_r d_k \hat{\phi}_k \mathbf{x}_k = \sum_r d_k v_{sk} \mathbf{x}_k = \begin{pmatrix} \hat{\mathbf{X}}^* \\ \hat{\mathbf{X}}^\circ \end{pmatrix}$$

where $\hat{\mathbf{X}}^* = \sum_s d_k \mathbf{x}_k^*$. An alternative for estimating the variance of the calibration estimator $\hat{Y}_W$ is to use v_{sk} instead of v_k in Proposition 11.1. That is, in (11.3)–(11.6), replace v_k by v_{sk} given by (10.2). This requires the individual value $\mathbf{x}_k^*$ for every $k \in s$. The v_{sk} entirely reflect a relation between the respondents, $k \in r$, and the sampled elements, $k \in s$, whereas the v_k depend on population totals. The variance estimate (11.3) will, in most cases, be roughly the same under either specification. The numerical differences are small in the empirical illustration reported in Section 11.5. $\quad\Box$

11.4. VARIANCE ESTIMATION FOR THE TWO-STEP A AND TWO-STEP B PROCEDURES

With minor modifications of Proposition 11.1, we obtain variance estimators for the two-step estimators $\hat{Y}_{W2A}$ and $\hat{Y}_{W2B}$. For two-step A, the first step calls for intermediate weights to be computed that are calibrated up to the level of the sample. They are

$$w_k^\circ = d_k v_k^\circ, \qquad v_k^\circ = 1 + (\boldsymbol{\lambda}_r^\circ)' \mathbf{x}_k^\circ \qquad (11.16)$$

where $(\boldsymbol{\lambda}_r^\circ)' = \left(\sum_s d_k \mathbf{x}_k^\circ - \sum_r d_k \mathbf{x}_k^\circ \right)' \left(\sum_r d_k \mathbf{x}_k^\circ (\mathbf{x}_k^\circ)' \right)^{-1}$. For example, in a one-way classification of the sampled elements, with $\mathbf{x}_k^\circ$ set equal to the classification vector $\boldsymbol{\gamma}_k = (\gamma_{1k}, \dots, \gamma_{pk}, \dots, \gamma_{Pk})'$, we have $w_k^\circ = d_k v_k^\circ$ with $v_k^\circ = \sum_{s_p} d_k / \sum_{r_p} d_k$ for all elements in the responding subset r_p of s_p. In the second step, $d_{\alpha k} = w_k^\circ = d_k v_k^\circ$ are used as initial weights in (6.14) to produce the final weights

$$w_{2Ak} = d_k v_k^\circ v_{2Ak}, \qquad v_{2Ak} = 1 + \boldsymbol{\lambda}_r' \mathbf{z}_k \qquad (11.17)$$

where $\boldsymbol{\lambda}_r' = \left(\mathbf{X} - \sum_r d_k v_k^\circ \mathbf{x}_k \right)' \left(\sum_r d_k v_k^\circ \mathbf{z}_k \mathbf{x}_k' \right)^{-1}$. The resulting two-step A calibration estimator is

$$\hat{Y}_{W2A} = \sum_r w_{2Ak} y_k. \qquad (11.18)$$

The variance estimator for $\hat{Y}_{\text{W2A}}$ is given by (11.3) if we replace v_k by v_k° in (11.4)–(11.6). Similarly to single-step, the justification relies on a substitution with the aid of proxies for the unknown influences. The equality $\hat{Y}_{\text{G2p}|\phi=\hat{\phi}} = \hat{Y}_{\text{W2A}}$ holds for $\phi_k = \hat{\phi}_k = v_k^{\circ}$, where v_k° is given by (11.16). To see this identity, denote by $g_{k\hat{\phi}}$ the value of $g_{k\phi}$ for $\phi_k = \hat{\phi}_k = v_k^{\circ}$. We then have

$$g_{k\hat{\phi}} = 1 + \left(\mathbf{X} - \sum_r d_k v_k^{\circ} \mathbf{x}_k \right)' \left(\sum_r d_k v_k^{\circ} \mathbf{z}_k \mathbf{x}_k' \right)^{-1} \mathbf{z}_k$$

and $d_k v_k^{\circ} g_{k\hat{\phi}} = w_{2Ak}$ given by (11.17). Therefore, for any fixed s and r,

$$\hat{Y}_{\text{G2p}|\phi=\hat{\phi}} = \sum_r d_k v_k^{\circ} g_{k\hat{\phi}} y_k = \sum_r w_{2Ak} y_k = \hat{Y}_{\text{W2A}}.$$

We obtain the proposed variance estimator for $\hat{Y}_{\text{W2A}}$ by substituting $\hat{\phi}_k = v_k^{\circ}$ for ϕ_k in (11.12). The proxies $\hat{\phi}_k = v_k^{\circ}$ have the calibration property

$$\sum_r d_k \hat{\phi}_k \mathbf{x}_k^{\circ} = \sum_r d_k v_k^{\circ} \mathbf{x}_k^{\circ} = \sum_s d_k \mathbf{x}_k^{\circ} = \hat{\mathbf{X}}^{\circ}.$$

They are controlled only by the information $\hat{\mathbf{X}}^{\circ}$, not by the information $\mathbf{X}^* = \sum_U \mathbf{x}_k^*$.

Two-step B uses the same intermediate weights as two-step A, $w_k^{\circ} = d_k v_k^{\circ}$, where v_k° is given by (11.16). The second step takes the $w_k^{\circ} = d_k v_k^{\circ}$ as initial weights in (6.14); the auxiliary vector is $\mathbf{x}_k = \mathbf{x}_k^*$, the information input is $\mathbf{X}^* = \sum_U \mathbf{x}_k^*$, and $\mathbf{z}_k^*$ is of the same dimension as $\mathbf{x}_k^*$. The resulting final weights are

$$w_{2Bk} = d_k v_k^{\circ} v_{2Bk}, \qquad v_{2Bk} = 1 + \boldsymbol{\lambda}' \mathbf{z}_k^* \qquad (11.19)$$

where $\boldsymbol{\lambda}_r' = \left(\mathbf{X}^* - \sum_r w_k^{\circ} \mathbf{x}_k^* \right)' \left(\sum_r w_k^{\circ} \mathbf{z}_k^* (\mathbf{x}_k^*)' \right)^{-1}$. The two-step B calibration estimator is

$$\hat{Y}_{\text{W2B}} = \sum_r w_{2Bk} y_k.$$

Although the weights w_{2Bk} of $\hat{Y}_{\text{W2B}}$ satisfy $\sum_r w_{2Bk} \mathbf{x}_k^* = \sum_U \mathbf{x}_k^* = \mathbf{X}^*$, they are less controlled than those of $\hat{Y}_{\text{W2A}}$. They fail to satisfy $\sum_r w_{2Bk} \mathbf{x}_k^{\circ} = \sum_s d_k \mathbf{x}_k^{\circ} = \hat{\mathbf{X}}^{\circ}$. The information carried by the moon vector $\mathbf{x}_k^{\circ}$ is not entirely wasted, however; it enters into the intermediate weights.

The variance estimator for $\hat{Y}_{\text{W2B}}$ is given by (11.3) in Proposition 11.1 if we replace v_k by v_k° in (11.4)–(11.6) and change the residuals into

$$\hat{e}_k^* = \hat{e}_k = y_k - (\mathbf{x}_k^*)' \left(\sum_r d_k v_k^{\circ} \mathbf{z}_k^* (\mathbf{x}_k^*)' \right)^{-1} \sum_r d_k v_k^{\circ} \mathbf{z}_k^* y_k.$$

Here, the residuals $\hat{e}_k$ remove the influence of $\mathbf{x}_k^*$ only, not that of $\mathbf{x}_k^{\circ}$, as was the case in two-step A.

The procedure is justified by the equality that we can establish between $\hat{Y}_{\text{W2B}}$ and a two-phase GREG estimator with the same information input, $\hat{Y}_{\text{G2p}} = \sum_r w_{\text{G2p}*k} y_k$, with weights given by

$$w_{\text{G2p}*k} = d_k \phi_k g_{k\phi}, \qquad g_{k\phi} = 1 + \left(\mathbf{X}^* - \sum_r d_k \phi_k \mathbf{x}_k^* \right)' \left(\sum_r d_k \phi_k \mathbf{z}_k^* (\mathbf{x}_k^*)' \right)^{-1} \mathbf{z}_k^*.$$
$$(11.20)$$

The equality $\hat{Y}_{G2p|\phi=\hat{\phi}} = \hat{Y}_{W2B}$ holds for $\phi_k = \hat{\phi}_k = v_k^\circ$, where v_k° is given by (11.16). The indicated variance estimator follows.

11.5. A SIMULATION STUDY OF THE VARIANCE ESTIMATION TECHNIQUE

Empirical studies throw light on the performance of the variance estimation technique in Proposition 11.1. Some evidence is provided in this section.

Given that there exists a method for variance estimation for the calibration estimator $\hat{Y}_W$, we look for it to accomplish two tasks:

(i) to give a reliable indication of the precision (through the variance or the standard error) of $\hat{Y}_W$;
(ii) to provide a basis for a valid confidence interval.

The interval for an approximately 95 % level of confidence is computed by the rule

$$\text{point estimate} \pm 1.96 \,(\text{variance estimate})^{1/2}.$$

It is based on the notion of an approximate normal distribution of $\hat{Y}_W$, over repeated realizations of samples and response sets. Although normality holds only approximately, this flaw is relatively minor. The most serious threat to the validity of the confidence interval is the bias of the point estimator $\hat{Y}_W$.

The confidence intervals have a frequentist interpretation. An interval has a 95 % confidence level if it includes the target value 95 % of the time, with respect to a long series of samples and response sets. Then we say that the interval is *valid*. It delivers the degree of confidence for which it was constructed.

When the point estimator is affected by bias, as it is to a greater or lesser extent in the presence of nonresponse, it must be remembered that (i) does not guarantee (ii). When $\hat{Y}_W$ is biased, the interval is off-centre. On average, the interval midpoint is not at the target value Y. The error rate, that is, the rate at which the interval will fail to include the target value Y, may considerably exceed the 5 % nominal rate, even if the variance estimator is unbiased or very nearly so. We shall see that objective (i) is reasonably well met, but that objective (ii) is in peril.

We examine once again the eight well-known estimators derived in Chapter 7 as special cases of the calibration estimator $\hat{Y}_W$ defined, in general form, by the weighting (6.14) with $d_{\alpha k} = d_k$. The eight estimators correspond to equally many different inputs of auxiliary information. The associated specifications ($\mathbf{x}_k$, $\mathbf{z}_k$) were listed in Table 10.1.

For seven of those eight estimators, the auxiliary vector is a star vector, $\mathbf{x}_k = \mathbf{x}_k^*$, so the whole information is at the population level. With the abbreviated names used earlier, seven are of this kind: EXP, PWA, RA, SEPRA, REG, SEPREG and TWOWAY. For one estimator, we have $\mathbf{x}_k = \mathbf{x}_k^\circ$, a moon vector, and the information is exclusively at the sample level. This estimator is

WC, where $\mathbf{x}_k = \mathbf{x}_k^{\circ} = \boldsymbol{\gamma}_k = (\gamma_{1k}, \ldots, \gamma_{pk}, \ldots, \gamma_{Pk})'$, the group identifier defined by (7.3). The eight estimators were studied by simulation in Section 7.7 with regard to their bias and variance.

Chapter 8 dealt with surveys where both kinds of information are available and used together in the estimation. Some information is at the sample level, some at the population level. Both kinds are combined to make calibration estimates. Of the three procedures in Section 8.1, we have emphasized single-step. With the auxiliary vector written as $\mathbf{x}_k = \begin{pmatrix} \mathbf{x}_k^* \\ \mathbf{x}_k^{\circ} \end{pmatrix}$ and $\mathbf{z}_k = \mathbf{x}_k$, this gives the estimator $\hat{Y}_W$ given by (8.1). Sections 8.2 and 8.3 dealt with one specific single-step estimator, $\hat{Y}_W = \hat{Y}_{SS:WC/REG}$. Given by (8.7), it is characterized by $\mathbf{x}_k^{\circ} = \boldsymbol{\gamma}_k = (\gamma_{1k}, \ldots, \gamma_{pk}, \ldots, \gamma_{Pk})'$, known for every $k \in s$, and $\mathbf{x}_k^* = (1, x_k)'$ with the known population total $(N, \sum_U x_k)'$. In addition, $\boldsymbol{\gamma}_k$ and x_k have known values for every element $k \in r$. We use the abbreviated name SS:WC/REG.

The conditions for the simulations in Sections 7.7 and 8.3 are kept intact. The experimental population is KYBOK, repeated SI samples are drawn, and the same two response distributions are used. We drew 10 000 SI samples from this population. For each sample, two response sets were obtained, one with the logit response distribution, the other with the increasing exponential response distribution. The conclusions that we can draw are limited by the restriction to only two response distributions.

Thus the nine estimators taking part in the simulation are EXP, PWA, RA, SEPRA, REG, SEPREG and TWOWAY (with population-level information only), WC (with sample-level information only), and SS:WC/REG (with combined sample-level and population-level information). As in Section 7.7, $\boldsymbol{\gamma}_k$ is the vector of dimension $P = 4$ that indicates type of clerical municipality and x_k is the square root of revenue advances.

Let $\hat{Y}_W$ be one of the nine estimators in the experiment, and let $\hat{V}(\hat{Y}_W)$ be the corresponding variance estimator, as given by (11.3). For every response set r_j, $j = 1, 2, \ldots, 10\,000$, we computed the point estimate $\hat{Y}_{W(j)}$ and the variance estimate $\hat{V}(\hat{Y}_{W(j)})$. Summary measures were computed by the following formulae:

(a) the simulation relative bias of the variance estimator $\hat{V}(\hat{Y}_W)$, given by

$$RB_{SIM}[\hat{V}(\hat{Y}_W)] = \frac{E_{SIM}[\hat{V}(\hat{Y}_W)] - V_{SIM}(\hat{Y}_W)}{V_{SIM}(\hat{Y}_W)} \qquad (11.21)$$

where

$$E_{SIM}[\hat{V}(\hat{Y}_W)] = \frac{1}{10\,000} \sum_{j=1}^{10\,000} \hat{V}(\hat{Y}_{W(j)})$$

is the simulation expectation of the variance estimator. As in Section 7.7,

$$V_{SIM}(\hat{Y}_W) = \frac{1}{9999} \sum_{j=1}^{10\,000} [\hat{Y}_{W(j)} - E_{SIM}(\hat{Y}_W)]^2$$

is the simulation variance of the point estimator $\hat{Y}_W$;

(b) the simulation coverage rates for a nominal 95 % confidence level, based on $\hat{Y}_W$ and its variance estimator $\hat{V}(\hat{Y}_W)$, and given by

$$CR_{SIM}\left[\hat{V}(\hat{Y}_W)\right] = \frac{1}{10\,000} \sum_{j=1}^{10\,000} I_{(j)} \qquad (11.22)$$

with

$$I_{(j)} = \begin{cases} 1 & \text{if } [a_{1j}, a_{2j}] \text{ contains } Y, \\ 0 & \text{otherwise} \end{cases}$$

where

$$a_{1j} = \hat{Y}_{W(j)} - 1.96[\hat{V}(\hat{Y}_{W(j)})]^{1/2}, \qquad a_{2j} = \hat{Y}_{W(j)} + 1.96[\hat{V}(\hat{Y}_{W(j)})]^{1/2}.$$

If we find that the simulation renders a small absolute value of $RB_{SIM}[\hat{V}(\hat{Y}_W)]$, say, not exceeding 5 % for any one estimator, it contributes evidence that the variance estimation can work satisfactorily.

For every estimator, the simulation should, ideally, produce a value of $CR_{SIM}[\hat{V}(\hat{Y}_W)]$ very close to 95 %. But the reality is different; $CR_{SIM}[\hat{V}(\hat{Y}_W)]$ will differ from 95 %, more or less. Several causes can be identified. Three of them are: the more or less extensive bias of the point estimator $\hat{Y}_W$; a certain bias of the corresponding variance estimator $\hat{V}(\hat{Y}_W)$; and the lack of a perfect normal distribution, over repeated response sets, of the estimator $\hat{Y}_W$. A realistic objective is to come 'reasonably close' to 95 %, at least for the 'better estimators', that is, those of the nine which receive better bias protection on account of stronger auxiliary information. (Both $RB_{SIM}[\hat{V}(\hat{Y}_W)]$ and $CR_{SIM}[\hat{V}(\hat{Y}_W)]$ are subject to some simulation error. Although 10 000 repetitions may be adequate for the exercise in this section, this number of repetitions will not in general suffice for full stability in either of the two quantities.)

Table 11.1 shows the simulation results, expressed as percentages for both quantities. The values for $RB_{SIM}(\hat{Y}_W)$ are taken from Tables 7.2 and 8.1. The results invite the following comments:

• The expansion estimator EXP is included strictly as a basis for comparison. It would hardly be considered in practice. Its unsatisfactory performance is expected. The bias is unacceptably high, especially in the point estimator. Some compensation occurs: despite a large bias in the point estimator, the coverage rate is still maintained at reasonably high levels by confidence intervals whose average length is inflated by the overestimation of the variance. In the following points we leave EXP out of consideration and comment on the other eight better-performing estimators.

• On the whole, the variance estimation is satisfactory. The bias of the variance estimator $\hat{V}(\hat{Y}_W)$ stays within reasonable bounds for all estimators. The objective of an absolute $RB_{SIM}[\hat{V}(\hat{Y}_W)]$ of less than 5 % is achieved, except for SEPREG. The tendency is for some positive bias for the weaker estimators, WC, PWA and RA, and for some negative bias for the stronger estimators, REG, SEPRA, SEPREG, TWOWAY and SS:WC/REG. The underestimation

Table 11.1 Simulation relative bias of $\hat{V}(\hat{Y}_W)$ (%), simulation coverage rate (%) and simulation relative bias of $\hat{Y}_W$ (%) for two response distributions and different point estimators. Samples of size $n = 300$ drawn by SI. Nonresponse rate 14%.

Response distribution	Point estimator	$RB_{SIM}[\hat{V}(\hat{Y}_W)]$ $\times 10^2$	$CR_{SIM}[\hat{V}(\hat{Y}_W)]$ $\times 10^2$	$RB_{SIM}(\hat{Y}_W)$ $\times 10^2$
Logit	Expansion (EXP)	5.4	94.2	5.0
	Weighting class (WC)	0.3	94.6	2.2
	Population weighting adjustment (PWA)	−1.2	93.7	2.2
	Ratio (RA)	2.5	93.8	2.5
	Separate ratio (SEPRA)	−3.4	93.6	0.7
	Regression (REG)	−3.5	92.8	−0.6
	Separate regression (SEPREG)	−12.1	91.9	−0.2
	Two-way classification (TWOWAY)	−4.1	92.3	0.5
	Single-step (SS:WC/REG)	−1.7	92.7	−0.6
Increasing exponential	Expansion (EXP)	6.2	85.9	9.3
	Weighting class (WC)	1.0	91.6	5.7
	Population weighting adjustment (PWA)	−0.4	86.1	5.7
	Ratio (RA)	2.6	90.5	3.9
	Separate ratio (SEPRA)	−2.3	90.6	−2.7
	Regression (REG)	−6.5	78.5	2.0
	Separate regression (SEPREG)	−12.3	90.6	−0.8
	Two-way classification (TWOWAY)	−4.7	92.6	0.5
	Single-step (SS:WC/REG)	−0.2	83.1	−2.4

of the variance conforms to a pattern mentioned in connection with the GREG estimator in Remark 4.1. Corrective action along lines mentioned in Remark 11.2 could be undertaken.

- Coverage rates substantially below the nominal 95% are unsatisfactory, but we know beforehand that they will happen, because of a pronounced nonresponse bias in some of the nine estimators. Under the circumstances, a coverage rate greater than 92% may be considered acceptable. Inadequate coverage rates occur when two conditions prevail: (a) there is bias, positive or negative, in the point estimator, so that $|RB_{SIM}(\hat{Y}_W)|$ is nonnegligible; and (b) there is negative bias in the variance estimator, so that $RB_{SIM}[\hat{V}(\hat{Y}_W)] < 0$. Of these, (a) is the principal cause of harm to the coverage rate. The problem is aggravated by (b), which causes the interval length to fall short on average. The 92% goal for the coverage rate is achieved under the logit response distribution for all estimators except SEPREG, which comes close. More interesting is to observe the results for the increasing exponential response distribution.

There are four out of the eight estimators for which $|RB_{\text{SIM}}(\hat{Y}_W)|$ is at least 2 % while at the same time $RB_{\text{SIM}}[\hat{V}(\hat{Y}_W)]$ is negative. They are PWA, REG, SEPRA and SS:WC/REG. Not surprisingly, they yield low coverage rates, ranging from 83.1 % to 90.6 %.

- The concern expressed in Remark 11.1 was exemplified by this simulation. For SEPREG, the results on $RB_{\text{SIM}}[\hat{V}(\hat{Y}_W)]$ and $CR_{\text{SIM}}[\hat{V}(\hat{Y}_W)]$ in Table 11.1 were computed following a substitution for negative values of v_k. For a few response sets containing one or more negative values of v_k, the variance estimator gave erratic results because of erratic residuals. Therefore, whenever negative v_k occurred for SEPREG, they were replaced in the variance estimator (but not in the point estimator) by the value 0.05, while the other v_k-values were kept intact. For all of the other estimators in Table 11.1, this substitution had no impact on the results, to the number of decimals reported.

The simulation indicates that although the variance can be estimated quite accurately, there is no guarantee that valid confidence intervals will follow. The bias of the point estimator is to blame. Even estimators that profit from relatively stronger information input have a bias so influential that the desired confidence level is not preserved. This is a message of serious consequence for all statistical inference from data affected by nonresponse.

Remark 11.5

We also studied the alternative variance estimate for $\hat{Y}_W$ mentioned in Remark 11.4: we replaced v_k in (11.3)–(11.6) by v_{sk} given by (10.2). The numerical differences in the resulting variance estimates were small and inconsequential.   □

We turn to the two-step A and two-step B procedures. For the information called WC/REG, the estimators $\hat{Y}_{2A:\text{WC/REG}}$ and $\hat{Y}_{2B:\text{WC/REG}}$ are given, respectively, by (8.9) and (8.11). Their respective variance estimators follow from Section 11.4. We studied them by simulation under the same conditions as for Table 11.1. The results are given in Table 11.2. Some overestimation of variance occurs. It is particularly pronounced for the increasing exponential response distribution, suggesting that $\hat{\phi}_k = v_k^\circ$ are less efficient proxies for variance estimation than $\hat{\phi}_k = v_k$ used

Table 11.2 Simulation relative bias of $\hat{V}(\hat{Y}_W)$ (%), simulation coverage rate (%) and simulation relative bias of $\hat{Y}_W$ (%) for two response distributions. Estimators by two-step A and two-step B. Samples of size $n = 300$ drawn by SI. Nonresponse rate 14 %.

Response distribution	Point estimator	$RB_{\text{SIM}}[\hat{V}(\hat{Y}_W)]$ $\times 10^2$	$CR_{\text{SIM}}[\hat{V}(\hat{Y}_W)]$ $\times 10^2$	$RB_{\text{SIM}}(\hat{Y}_W)$ $\times 10^2$
Logit	Two-step A (2A:WC/REG)	3.8	93.0	−0.6
	Two-step B (2B:WC/REG)	4.9	92.7	−0.8
Increasing	Two-step A (2A:WC/REG)	22.8	87.8	−2.3
exponential	Two-step B (2B:WC/REG)	24.7	83.7	−3.0

for SS:WC/REG in Table 11.1. For the increasing exponential response distribution, it is the considerable negative bias of the point estimator that is to blame for coverage rates that drop well below 90 %. This is particularly noticeable for two-step B, even though the interval width is overstated on average, as a result of the positive bias in the variance estimator.

11.6. COMPUTATIONAL ASPECTS OF POINT AND VARIANCE ESTIMATION

Two of the basic ideas in statistical estimation theory are (i) point estimation and (ii) variance estimation with an accompanying confidence interval computation. In particular, for calibration estimation in the presence of nonresponse, the two steps are (i) computation of the calibrated weights and the point estimator, as described in Chapter 6, and (ii) computation of a corresponding variance estimator, which may rely on the methods of the present chapter.

The national statistical agencies in a number of countries have developed software for estimation procedures that use auxiliary information. Among such countries are Canada, France, Italy, Sweden, the Netherlands and the United States. In general, such software is capable, possibly with some modification, of carrying out point estimation according to the theory in Chapters 6–8. Some can also handle the corresponding variance estimation. Because survey environments vary, different countries will approach the development of software from different angles. The sampling designs in popular use differ between countries. In a number of European countries, and particularly in Scandinavia, stratified sampling from the population register is a dominating design for surveys on households and individuals. In North America, such surveys are typically handled by sampling in two or more stages, and this is reflected in the variance estimation.

Software can be more or less efficient in the computation of the calibrated weights. The matrix inversion in (6.10) and (6.14) constitutes a potential trouble spot, especially when $\mathbf{x}_k$ includes one or more quantitative variables. If the computation proceeds in the direct manner, with an $\mathbf{x}_k$-vector that is dummy coded for each categorical variable, the matrix to invert becomes a 'sparse matrix', containing a number of zeros. Matrix inversion and other parts of the weight computation may be slow. With computational ingenuity, this inconvenience is avoided, as in several existing software programs.

Some programs have the capacity to compute point estimates and corresponding variance estimates for more complex parameters, constructed as functions of totals. Consider the parameter $\psi = f(Y_1, \ldots, Y_m, \ldots, Y_M)$, where f is a specified function of the M population totals $Y_1, \ldots, Y_m, \ldots, Y_M$. In particular, rational functions are of interest in many surveys. (A rational function is one that is limited to use of the four basic algebraic rules, addition, subtraction, multiplication and division.) An example of such a parameter is a difference of ratios, $\psi = Y_1/Y_2 - Y_3/Y_4$. For example, given any parameter defined as a rational function of totals, CLAN97 will compute (i) the point estimate determined as $\hat{\psi} = f(\hat{Y}_{1\mathrm{W}}, \ldots, \hat{Y}_{m\mathrm{W}}, \ldots, \hat{Y}_{M\mathrm{W}})$, where $\hat{Y}_{1\mathrm{W}}, \ldots, \hat{Y}_{m\mathrm{W}}, \ldots, \hat{Y}_{M\mathrm{W}}$ are the respective calibration estimates of the M totals (population totals or domain totals), and (ii) the corresponding variance estimate.

APPENDIX: PROPERTIES OF THE TWO-PHASE GREG ESTIMATOR

The objective in this appendix is to derive the basic properties of the two-phase estimator $\hat{Y}_{\mathrm{G2p}}$ given by (11.10). We use the notation of Section 11.2. Because of the complex nonlinear form of $\hat{Y}_{\mathrm{G2p}}$, the expected value and the variance cannot be stated exactly. But we can rely on the theory of two-phase sampling, as presented in, for example, Särndal et al. (1992). First we derive the linearized statistic, $\hat{Y}_{\mathrm{G2p,lin}}$, corresponding to $\hat{Y}_{\mathrm{G2p}}$. Exact expressions for the bias and variance of $\hat{Y}_{\mathrm{G2p,lin}}$ are easily obtained; the bias is exactly zero. Because $\hat{Y}_{\mathrm{G2p}} \approx \hat{Y}_{\mathrm{G2p,lin}}$, we can take these expressions as close approximations to the corresponding moments of $\hat{Y}_{\mathrm{G2p}}$. The variance of $\hat{Y}_{\mathrm{G2p,lin}}$ is obtained as the sum of two components V_{SAM}, the sampling variance, and V_{NR}, the additional variance caused by the nonresponse. As one would expect, V_{NR} is zero if all elements respond with certainty, that is, if $\phi_k = 1/\theta_k = 1$ for all k. Estimated variance components are derived by a straightforward transition to the sample-based analogues of V_{SAM} and V_{NR}. We thus obtain the variance estimator stated by (11.12)–(11.15). The following proposition and its proof provide the details.

Proposition A11.1

By linearization, $\hat{Y}_{\mathrm{G2p}} \approx \hat{Y}_{\mathrm{G2p,lin}}$, where the linearized statistic is

$$\hat{Y}_{\mathrm{G2p,lin}} = \sum_r d_k \phi_k (y_k - \mathbf{x}_k' \mathbf{B}_U) + \left(\sum_U \mathbf{x}_k \right)' \mathbf{B}_U + \left(\sum_s d_k \mathbf{x}_k^{\circ} - \sum_U \mathbf{x}_k^{\circ} \right)' \mathbf{B}_U^{\circ}$$

in which $\mathbf{x}_k = \begin{pmatrix} \mathbf{x}_k^* \\ \mathbf{x}_k^{\circ} \end{pmatrix}$, $\mathbf{z}_k$ is a vector of the same dimension as $\mathbf{x}_k$, and

$$\mathbf{B}_U = \begin{pmatrix} \mathbf{B}_U^* \\ \mathbf{B}_U^{\circ} \end{pmatrix} = \left(\sum_U \mathbf{z}_k \mathbf{x}_k' \right)^{-1} \left(\sum_U \mathbf{z}_k y_k \right). \qquad (A11.1)$$

The variance of $\hat{Y}_{\mathrm{G2p}}$ is approximated by

$$V(\hat{Y}_{\mathrm{G2p}}) \approx V(\hat{Y}_{\mathrm{G2p,lin}}) = V_{\mathrm{SAM}} + V_{\mathrm{NR}} \qquad (A11.2)$$

where

$$V_{\mathrm{SAM}} = \sum\sum_U \left(\frac{d_k d_\ell}{d_{k\ell}} - 1 \right) e_k^* e_\ell^*, \qquad V_{\mathrm{NR}} = E_p \left[\sum_s (\phi_k - 1)(d_k e_k)^2 \right]$$

with the residuals

$$e_k^* = y_k - (\mathbf{x}_k^*)' \mathbf{B}_U^*, \qquad e_k = y_k - \mathbf{x}_k' \mathbf{B}_U = y_k - (\mathbf{x}_k^*)' \mathbf{B}_U^* - (\mathbf{x}_k^{\circ})' \mathbf{B}_U^{\circ}.$$

Here, E_p is the expected value operator with respect to the sample selection (phase 1). The estimated variance is given by (11.12), where the components $\hat{V}_{\mathrm{SAM}}$ and $\hat{V}_{\mathrm{NR}}$ are given, respectively, by (11.13) and (11.14). □

Proof. We can write (11.10) as $\hat{Y}_{G2p} = \sum_r d_k \phi_k (y_k - \mathbf{x}'_k \mathbf{B}_{r;d\phi}) + \mathbf{X}' \mathbf{B}_{r;d\phi}$ where $\mathbf{X} = \begin{pmatrix} \mathbf{X}^* \\ \hat{\mathbf{X}}^\circ \end{pmatrix}$ and

$$\mathbf{B}_{r;d\phi} = \begin{pmatrix} \mathbf{B}^*_{r;d\phi} \\ \mathbf{B}^\circ_{r;d\phi} \end{pmatrix} = \left(\sum_r d_k \phi_k \mathbf{z}_k \mathbf{x}'_k \right)^{-1} \left(\sum_r d_k \phi_k \mathbf{z}_k y_k \right).$$

Over repeated realizations r and s, the random vector $\mathbf{B}_{r;d\phi}$ fluctuates around its population analogue $\mathbf{B}_U$ given by (A11.1). Using $\mathbf{B}_{r;d\phi} = \mathbf{B}_U + (\mathbf{B}_{r;d\phi} - \mathbf{B}_U)$, we can express $\hat{Y}_{G2p}$ as the sum of a main term, the linearized statistic, and a lower-order term involving the small difference $\mathbf{B}_{r;d\phi} - \mathbf{B}_U$. We have

$$\hat{Y}_{G2p} = \hat{Y}_{G2p,\text{lin}} - R$$

where the linear term is

$$\hat{Y}_{G2p,\text{lin}} = \sum_r d_k \phi_k (y_k - \mathbf{x}'_k \mathbf{B}_U) + \left(\sum_U \mathbf{x}_k \right)' \mathbf{B}_U + \left(\sum_s d_k \mathbf{x}^\circ_k - \sum_U \mathbf{x}^\circ_k \right)' \mathbf{B}^\circ_U$$

and the remainder term is $R = \left(\sum_r d_k \phi_k \mathbf{x}_k - \mathbf{X} \right)' (\mathbf{B}_{r;d\phi} - \mathbf{B}_U)$. Both $\hat{Y}_{G2p,\text{lin}} - Y$ and R are near zero with high probability in large samples and response sets, whereas R is of lower order of importance. Thus, to close approximation, $\hat{Y}_{G2p} \approx \hat{Y}_{G2p,\text{lin}}$, and $V(\hat{Y}_{G2p}) \approx V(\hat{Y}_{G2p,\text{lin}})$. Note that $\hat{Y}_{G2p,\text{lin}}$ is a statistic (random, through its dependence on s and r) with the desired expectation, $E_{pq}(\hat{Y}_{G2p,\text{lin}}) = Y$, but it is not an estimator even though the ϕ_k are known, because it contains the unknown population vector $\mathbf{B}_U$. We derive the variance of $\hat{Y}_{G2p,\text{lin}}$ (which closely approximates the variance of $\hat{Y}_{G2p}$) by use of the well-known rule 'variance of the conditional expectation plus expectation of the conditional variance', where 'conditional' means 'given the sample s'. These two components are

$$V_p E_q (\hat{Y}_{G2p,\text{lin}}) = V_p \left(\sum_s d_k e^*_k \right), \qquad E_p V_q (\hat{Y}_{G2p,\text{lin}}) = E_p V_q \left(\sum_r d_k \phi_k e_k \right).$$

From these expressions follow the two variance components in (A11.2). The final step is to transform (A11.2) into an estimated variance: replace the double sums over U by the appropriately weighted double sum over r, and replace the unknowns in e^*_k and e_k by their corresponding sample-based quantities, $\hat{e}^*_k$ and $\hat{e}_{k\phi}$, given at the end of Section 11.2. We have thus obtained the variance estimator defined by (11.12)–(11.15).   □

CHAPTER 12

Imputation

12.1. WHAT IS IMPUTATION?

Imputation is the procedure whereby missing values on one or more study variables are 'filled in' with substitutes. These substitutes can be constructed in a variety of ways, reviewed later in this chapter. Imputed values are by definition artificial. They contain error. *Imputation error* is similar to measurement error (as when a respondent provides an erroneous value for an item on a questionnaire) in that the value recorded in the data file is not the 'true value', except in truly exceptional circumstances. But unlike measurement error, imputation error occurs 'by construction', since the statistician inserts values known to be more or less wrong.

Another type of construction of artificial values, practised by some statistical agencies, is *mass imputation*. In this procedure, values are imputed not only for the sampled elements, but also for all nonobserved elements in the population. Mass imputation is not discussed in this book.

Imputed values can be classified into three major categories:

(i)   values constructed with the aid of a statistical prediction rule;
(ii) values observed not for the nonresponding elements themselves, but for (similar) responding elements;
(iii) values constructed by expert opinion or 'best possible judgement'.

Categories (i) and (ii) can be termed *statistical rules*, because they use common statistical techniques to produce substitute values. Category (i) is often based on an assumed relationship between variables, as in regression prediction. Category (ii) methods can be described as *donor-based*, in that the imputed value is 'borrowed' from an observed element considered, on good statistical grounds, to be 'similar' or, better still, 'very similar'. Category (iii) methods rely heavily on expert skill and knowledge about the particular element requiring imputation.

By another often-used distinction, imputed values are either *deterministic* (when repeating the imputation procedure would yield exactly the same imputed values) or *random* (when repeating the procedure would, barring pure chance, yield different imputed values). Regression imputation is an example of a

Estimation in Surveys with Nonresponse   C.-E. Särndal and S. Lundström
© 2005 John Wiley & Sons, Ltd

deterministic rule, whereas an example of a random rule is to impute the value of a randomly selected observed element. This procedure is often called 'hot deck imputation'.

Categories (i), (ii) and (iii) are discussed in more detail later in this chapter.

Many statisticians, including methodologists and subject-matter specialists, regard imputation with a somewhat critical eye. It goes against statistical common sense to compute reliable statistics with the aid of values known at the outset to be more or less wrong. On the other hand, the practical reasons for imputation are sometimes compelling.

There is no strong evidence that careful imputation does any more harm to survey estimates than certain other steps in the statistical production process. As is the case with weighting, one hopes that imputation will lead to estimates with, primarily, a small bias and, secondarily, a small variance.

Imputation may be preferable to weighting in some instances. An example is when the population is highly skewed, as is the case in many business surveys. Expert judgement may then be a preferred way to get a 'reasonably close' imputed value for a large, influential nonresponding element. Weighting may be less appealing for such an element. Weight computation always involves some form of smoothing, by reference to 'similar elements'. This is not ideal for a very large and unique element.

In general, sound imputation requires considerable professional skill and care. The imputed values must be close substitutes for the unobserved values. It is good policy to review and continuously upgrade the imputation methods used in a survey.

A distinction is often made between *imputed value* and *derived value*. An imputed value is a constructed value inserted in a data file in lieu of one that is missing. It is obtained by other means than observing the element in question and is unlikely to be exact or 'true'. A derived value, on the other hand, may also be inserted in a data file, but it is derived by an arithmetical operation for all elements in a sample or in the population. Derived values are usually considered 'true values'.

12.2. TERMINOLOGY

A large survey normally involves many study variables. They may be affected to varying extents by item nonresponse. In addition, there is usually also unit non-response. The concepts of item nonresponse and unit nonresponse were defined in Section 2.2.

The methods at our disposal to treat the nonresponse are weighting and imputation. Theoretically at least, this opens up three probabilities: we can rely strictly on the former, strictly on the latter, or on a combination of the two. Practical reasons may give preference to one of these three modes of operation.

Full weighting is the approach where weighting is practised one study variable at a time. There is no imputation. This approach calls for one set of weights to be computed for every study variable in the survey.

It is an indisputable advantage if all variables can be treated in a uniform manner in the production of statistics, with a common weighting applied to all study variables. Imputation makes this possible.

An often-mentioned reason for imputation is that it yields a *rectangular data set*. The desirability of this is particularly evident in a large survey involving multiple *y*-variables. Each record (each element) defines a row in a data matrix with *I* columns, where *I* is the number of *y*-variables. Before imputation, the data matrix contains a number of 'holes' caused by missing *y*-values. After imputation, every record has *I* recorded values, some observed, some imputed. The data set has become rectangular.

There are two frequently used approaches to imputation, both resulting in rectangular data matrices. They are full imputation and the combination of weighting and imputation.

Full imputation means that imputation is used to treat both item nonresponse and unit nonresponse. That is, we impute all missing values for all elements having one or more *y*-value missing. The resulting completed rectangular data matrix has dimension $n \times I$, where n is the sample size. There is no nonresponse weight adjustment.

Weighting and imputation combined, or the *combined approach* for short, implies that imputation is restricted to the item nonresponse. We impute for the m elements having at least one but not all *y*-values missing. The resulting rectangular data matrix has dimension $m \times I$. An appropriate weighting is then applied to compensate for the unit nonresponse. We consider the combined approach to be the principal alternative, in agreement with practice in many national statistical agencies.

If called for, both approaches will also make appropriate use of the sampling weights or weighting that incorporates available auxiliary information, as in the calibration approach.

Example 12.1. Some issues arising with imputation

The following simple example illustrates some of the terminology. Suppose that the survey has only one *y*-variable, of the continuous kind, and that every missing value is imputed by the average of the respondent *y*-values. The data set after imputation, called the completed data set, will then consist of m actually observed values, y_k, for $k \in r$, and $n - m$ imputed values, all of which are equal to $\bar{y}_r = \sum_r y_k/m$, where r is the response set of size m. As is intuitively clear, the method is inefficient and not recommended in a survey with high requirements on quality. It does achieve the objective of a completed data set. It is easy to identify several shortcomings of this data set. Usually, neither the central tendency nor the variance of these data will agree with what is expected of a data set with complete response. The variance will be unnaturally small, because $n - m$ missing values have been imputed by one and the same value, the respondent mean $\bar{y}_r$. Also, the central tendency of these data will often not reflect the true central tendency of the *y*-variable. If, for example, large *y*-value elements respond less often than small *y*-value elements, then the average for the completed data set is likely to fall short of the mean of a data set with complete response. If instead we impute the value of a randomly selected responding element (the 'donor element'), then the variability of the completed data set will start to look more 'normal'. But the central tendency will have the same shortcoming as in imputation by the respondent mean. Also, when the donor is randomly selected, we run the risk of imputing the value of a very large element by that of a very small element. The desired 'closeness to the truth' of the imputed value may be severely compromised by this simplistic procedure. The resulting estimates risk being severely biased.   □

12.3. MULTIPLE STUDY VARIABLES

As before, U and s denote, respectively, the target population and the probability sample drawn from U. The sampling weights are $d_k = 1/\pi_k$. We need additional notation to accommodate features of a survey involving multiple study variables, which may be affected to varying extents by item nonresponse. Consider a survey with I study variables (or items). We can focus on one of these variables, say, the ith variable, which we denote simply by y, whose value for element k is y_k. (For the $I - 1$ other study variables we need no specific notation.) The objective is to estimate the total $Y = \sum_U y_k$. If y_k is observed for every element $k \in s$, we say that there is *complete response* for variable y. In general, the variable y is affected both by unit nonresponse and by item nonresponse.

In line with the terminology in Section 2.2, r denotes the response set for the survey. This subset of the sample s consists of all elements having responded to one or more of the I study variables. An element k having responded to none of the I study variables belongs to the unit nonresponse set $s - r$. The set of elements having responded to the particular study variable y is denoted by r_i and is called the item response set. That is, the value y_k is observed for $k \in r_i$, where $r_i \subseteq r \subseteq s$. If y_k is missing and treated by imputation, let us denote the imputed value by $\hat{y}_k$. The notation is general; $\hat{y}_k$ denotes a value that may have been obtained by any one of the standard methods. More than one imputation method may be used in the same survey. That is, not all of the imputed values $\hat{y}_k$ may be the result of one and the same method. When $r_i = r$, there is no item nonresponse for the variable y. If $r = s$, the survey has no unit nonresponse, but y and the other $I - 1$ study variables may be affected to various extents by item nonresponse. Then $r_i - r$ may be nonempty for all i.

12.4. THE FULL IMPUTATION APPROACH

There exist many imputation methods. Some common methods are reviewed later. Each method has variations, because a minor modification of a well-established method may be needed to suit the requirements of a particular survey. In this section and the next, we discuss imputation within a framework sufficiently general to cover the various imputation methods in common use.

In the full imputation approach, we impute for all values y_k that are missing, whether by unit nonresponse or by item nonresponse. The resulting *completed data set* is the set of values $\{y_{\bullet k} : k \in s\}$, where

$$y_{\bullet k} = \begin{cases} y_k & \text{for } k \in r_i, \\ \hat{y}_k & \text{for } k \in s - r_i. \end{cases} \tag{12.1}$$

That is, $y_{\bullet k}$ stands for the observed value y_k when k is responding, and for the imputed value $\hat{y}_k$ when y_k is missing. All missing values have been replaced by surrogates. The result is a rectangular data matrix. At this point, $\hat{y}_k$ denotes a value that may have been constructed by any one of the imputation methods reviewed later in the chapter.

Traditional descriptive statistics, mean, variance and others, can be computed from the completed data set. For example, the mean of the completed data set is $\bar{y}_{\bullet s} = \sum_s y_{\bullet k}/n$. By contrast, the mean that would have been computed in the case of complete response is $\bar{y}_s = \sum_s y_k/n$. Both means are based on n values, but they will differ to an unknown extent. Similarly, we can compute a variance and other standard statistics from the completed data set. They also differ from their counterparts for a hypothetical data set consisting entirely of observed values.

The objective is to estimate the population total $Y = \sum_U y_k$ of the study variable y. In current estimation practice, imputed data are treated as real, observed data, at least when it comes to point estimation. One pretends that imputed values are 'as good as' true observations. This outlook invites the use of exactly the same estimation method as in the ideal case of complete response. We employ the 'standard estimator formula' and simply compute it on the completed data set. Consequently, element k receives the same weight whether the value recorded in the completed data file is a true observation, y_k, or an imputed value, $\hat{y}_k$. This is worth noting, because some would argue that imputed values and truly observed values should receive different treatment in the weighting process.

We call *complete response estimator* the one that would be used had y_k been observed for all sampled elements. This estimator has a well-defined formula. After imputation, we can compute this formula on the completed data set (12.1). The result is called the *imputed estimator*. The weighting system of the complete response estimator is used unchanged. The completed data set (12.1) simply replaces the desired (but unavailable) data set consisting entirely of real observations.

If the Horwitz–Thompson estimator, $\hat{Y}_{\mathrm{HT}} = \sum_s d_k y_k$, is designated as the complete response estimator, then the imputed counterpart is

$$\hat{Y}_{\mathrm{IHT}} = \sum_s d_k y_{\bullet k} = \sum_{r_i} d_k y_k + \sum_{s-r_i} d_k \hat{y}_k. \tag{12.2}$$

If the complete response estimator is the GREG estimator $\hat{Y}_{\mathrm{GREG}} = \sum_s d_k g_k y_k$, the weights are $d_k g_k$ for $k \in s$, as given by (4.10). The information input is $\sum_U \mathbf{x}_k^*$. The imputed counterpart is

$$\hat{Y}_{\mathrm{IGREG}} = \sum_s d_k g_k y_{\bullet k} = \sum_{r_i} d_k g_k y_k + \sum_{s-r_i} d_k g_k \hat{y}_k. \tag{12.3}$$

In current practice, point estimation following imputation is thus extremely simple, since the weights remain unchanged. By contrast, variance estimation is a more intricate issue.

We shall not pursue the full imputation approach in more detail. The combined approach, favoured in a majority of statistical agencies, is our main alternative.

12.5. THE COMBINED APPROACH

The widely employed combined approach uses imputation for the item nonresponse and weighting to compensate for the unit nonresponse. We develop the reasoning under the single-step calibration procedure defined in Section 8.1. The auxiliary vector is $\mathbf{x}_k = \begin{pmatrix} \mathbf{x}_k^* \\ \mathbf{x}_k^\circ \end{pmatrix}$. With the corresponding information $\mathbf{X} = \begin{pmatrix} \mathbf{X}^* \\ \hat{\mathbf{X}}^\circ \end{pmatrix}$, we can compute the single-step calibrated weights for $k \in r$ in accordance with (6.14) with $d_{\alpha k} = d_k$. They are

$$w_k = d_k v_k, \qquad v_k = 1 + \left(\mathbf{X} - \sum_r d_k \mathbf{x}_k \right)' \left(\sum_r d_k \mathbf{z}_k \mathbf{x}_k' \right)^{-1} \mathbf{z}_k. \qquad (12.4)$$

The weights $d_k v_k$ have built-in compensation for the unit nonresponse and have the desired calibration property $\sum_r d_k v_k \mathbf{x}_k = \mathbf{X}$. They are appropriate if the variable y is affected by unit nonresponse only, that is, when $r_i = r$. The appropriately weighted calibration estimator for the total $Y = \sum_U y_k$ is then

$$\hat{Y}_W = \sum_r d_k v_k y_k. \qquad (12.5)$$

A variance estimation procedure for $\hat{Y}_W$ is obtainable from Proposition 11.1.

Most likely, the variable y is affected both by item nonresponse and by unit nonresponse. Then the values y_k are available only for $k \in r_i \subset r \subset s$. The combined approach calls for an imputation step followed by a weighting step. We first impute for the item nonresponse elements, $k \in r - r_i$, so as to create a full rectangular data matrix with values specified for every study variable and for every element k in the response set r. (By contrast, in full imputation, as defined by (12.1), we also impute for $k \in s - r$.) The completed data for variable y are $\{y_{\bullet k} : k \in r\}$, where

$$y_{\bullet k} = \begin{cases} y_k & \text{for } k \in r_i, \\ \hat{y}_k & \text{for } k \in r - r_i \end{cases} \qquad (12.6)$$

that is, $y_{\bullet k}$ is the observed value y_k when k responds to this item, and otherwise $y_{\bullet k}$ is the imputed value $\hat{y}_k$. The method used to construct $\hat{y}_k$ may be any one of those discussed in Sections 12.7 and 12.8. We estimate the total $Y = \sum_U y_k$ by summing the appropriately weighted values $y_{\bullet k}$ over the response set r. The combined approach gives the *item imputed calibration estimator*

$$\hat{Y}_{IW} = \sum_r d_k v_k y_{\bullet k} = \sum_{r_i} d_k v_k y_k + \sum_{r-r_i} d_k v_k \hat{y}_k \qquad (12.7)$$

where the calibrated weights are $d_k v_k$, with v_k given by (12.4). We shall examine this estimator further, but note first that full weighting is another option.

12.6. THE FULL WEIGHTING APPROACH

An alternative to the combined approach is a complete reliance on weighting. No imputation is used. The estimator is computed as a sum of appropriately weighted responses y_k over the item response set r_i. To this end we use the calibrated weights given for $k \in r_i$ by

$$w_k = d_k v_{ik}, \qquad v_{ik} = 1 + \left(\mathbf{X} - \sum_{r_i} d_k \mathbf{x}_k \right)' \left(\sum_{r_i} d_k \mathbf{z}_k \mathbf{x}_k' \right)^{-1} \mathbf{z}_k. \qquad (12.8)$$

If $r_i = r$ in (12.8), we revert to the formula (12.4) appropriate for the case when there is only unit nonresponse. The weights $d_k v_{ik}$ compensate for both unit non-response and item nonresponse. They expand directly, so to speak, from the item response set r_i to the sample s, bypassing r. They have the calibration property $\sum_{r_i} d_k v_{ik} \mathbf{x}_k = \mathbf{X}$. The resulting *fully weighted calibration estimator* of $Y = \sum_U y_k$ is

$$\hat{Y}_{\text{FW}} = \sum_{r_i} d_k v_{ik} y_k \qquad (12.9)$$

with v_{ik} given by (12.8). If all r_i are different, the procedure requires different weights for every study variable. This may be viewed as less appealing in practice. One reason is that one may not wish to burden the survey data file with as many weight sets as there are study variables in the survey.

The item imputed estimator $\hat{Y}_{\text{IW}}$ given by (12.7) and the fully weighted estimator $\hat{Y}_{\text{FW}}$ given by (12.9) use different weight systems (unless $r_i = r$), but they do have one thing in common: the weights are computed on the same input of auxiliary information, $\mathbf{X} = \begin{pmatrix} \mathbf{X}^* \\ \hat{\mathbf{X}}^\circ \end{pmatrix}$. The weights $d_k v_{ik}$ in (12.9) are fewer in number, numerically greater on average, and create a stronger expansion than the weights $d_k v_k$ in (12.7). An important point is that (12.9) and (12.7) are identical for some types of imputation. This happens notably for a case of regression imputation, as discussed in Chapter 13.

Remark 12.1

Another instance where a fully weighted procedure is inconvenient occurs for a cross-tabulation. For example, consider a health survey where 'state of health' (categorical variable A) is cross-classified with 'type of professional activity' (categorical variable B). Consider the attribute defined by one particular category of A and one particular category of B. We wish to estimate the number of persons in the population with the cross-classified attribute in question. The dichotomous study variable y has the value $y_k = 1$ if person k has the attribute and $y_k = 0$ if not. The target parameter is $Y = \sum_U y_k$, the number in the population with the attribute.

In the fully weighted procedure with weights defined by (12.8), the item response set r_i is the set of elements having responded both to A (by indicating one of the fully exhaustive categories of A) and to B (by indicating one of the fully exhaustive categories of B). Now, other cross-classifications may be of interest in the survey, say 'state of health' (variable A) crossed with 'extent of physical activity' (variable C). To estimate the number of elements in a cell of that cross-classification, a new set r_i must be identified and used as a basis for computing the weights (12.8). In other words, in the fully weighted approach, every new cross-classification of interest requires a new set of calibrated weights. This may be seen as inconvenient. The combined approach is often preferable. □

12.7. IMPUTATION BY STATISTICAL RULES

We now examine a few commonly used rules to produce the imputed values $\hat{y}_k$, with an emphasis on the approach that combines weighting and imputation. For an in-depth discussion of imputation, the reader is referred to specialized treatises.

It is necessary to distinguish imputation by a statistical rule from special imputation. Methods in the former category are derived from statistical prediction arguments. These methods handle the bulk of the imputation effort needed in a survey. Special imputation, by expert judgement or by the use of historical data, is usually reserved for a selected few but highly influential elements.

Commonly used categories of statistical imputation rules are regression imputation, nearest neighbour imputation and hot deck imputation. They are broad categories; the basic ideas for each of them are described later in this section. Simple special cases of regression imputation are respondent mean imputation and ratio imputation.

Regression imputation and nearest neighbour imputation require auxiliary information. They give *deterministic* imputations. The auxiliary vector used to impute is denoted as before by $\mathbf{x}_k = (x_{1k}, \ldots, x_{jk}, \ldots, x_{Jk})'$ and is composed of the values of one or more *imputation variables*. We assume that the vector value $\mathbf{x}_k$ is known for every $k \in s$. When $\mathbf{x}_k$ is univariate, we write simply $\mathbf{x}_k = x_k$.

The imputation vector $\mathbf{x}_k$ is instrumental in producing the imputed values $\hat{y}_k$. If the imputation variable(s) in $\mathbf{x}_k$ are strong predictors for the imputed variable y, we can expect 'close imputations'. The imputation error for element k, $\hat{y}_k - y_k$, should be small.

Nearest neighbour and hot deck are *donor-based* methods, implying that the imputed value is one that was actually observed, although for a different element. This provides an assurance that the imputed value is one that can occur; it cannot be an impossible value. Hot deck, as described below, is a *random* imputation method, while nearest neighbour is *deterministic*.

Imputation by a statistical rule is often carried out mechanically, for a possibly large number of missing values, using existing computer software. An example is Statistics Canada's Generalised Editing and Imputation System software. Such mechanical imputation is often performed within *imputation groups*. These groups have to be identified at the outset. An imputation group is one deemed to consist of 'similar elements'.

Practitioners often impute according to a *hierarchy of methods*. A stronger method, one that is likely to produce 'closer' imputations, is first applied within one group of nonrespondents. Then, if the auxiliary information required for the preferred imputation method is not at hand for all elements, the second strongest method is applied within the next group, and so on.

Imputation by a statistical rule is motivated by the statistician's perception of a strong relationship between the study variable, y, and the imputation vector $\mathbf{x}$. We have observed values y_k only for the set r_i, the item response set for y. Thus imputed values are computed in the combined approach for $k \in r - r_i$, using the observed values y_k for $k \in r_i$ and other information.

Regression imputation

In the *regression imputation* method, the imputed value for a missing value y_k is

$$\hat{y}_k = \mathbf{x}'_k \hat{\boldsymbol{\beta}}_i \tag{12.10}$$

where

$$\hat{\boldsymbol{\beta}}_i = \left(\sum_{r_i} a_k \mathbf{x}_k \mathbf{x}'_k \right)^{-1} \sum_{r_i} a_k \mathbf{x}_k y_k. \tag{12.11}$$

The regression coefficient vector $\hat{\boldsymbol{\beta}}_i$ is the result of a multiple regression fit using the data $(y_k, \mathbf{x}_k)$ available for $k \in r_i$, and weighted with suitably specified values a_k. In the most straightforward case, all a_k are equal.

In the special case of a simple linear regression with an intercept, we have $\mathbf{x}_k = (1, x_k)'$, and the imputed value is $\hat{y}_k = \bar{y}_{r_i;a} - (x_k - \bar{x}_{r_i;a}) B_{r_i;a}$, where $\bar{y}_{r_i;a} = \sum_{r_i} a_k y_k / \sum_{r_i} a_k$, $\bar{x}_{r_i;a}$ is analogously defined, and $B_{r_i;a} = \sum_{r_i} a_k (x_k - \bar{x}_{r_i;a})(y_k - \bar{y}_{r_i;a}) / \sum_{r_i} a_k (x_k - \bar{x}_{r_i;a})^2$.

Two important special cases are *ratio imputation* and *respondent mean imputation*.

Ratio imputation and respondent mean imputation

When $\mathbf{x}_k = x_k$ is an always positive, one-dimensional imputation variable, and $a_k = 1/x_k$, the imputed value (12.10) becomes $\hat{y}_k = x_k \hat{\beta}_i$, with $\hat{\beta}_i = \sum_{r_i} y_k / \sum_{r_i} x_k$. This *ratio imputation* rule is often used when the same variable is measured on two different occasions in a repeated survey. Then y denotes the study variable at the present survey occasion, and x is the same variable at the preceding occasion. To illustrate, if y and x represent 'gross business income' at the two occasions, then the 'current ratio' $\hat{\beta}_i$ measures the change in the level of business income between the two occasions.

In particular, when $x_k = a_k = 1$ for all k, the imputed value (12.10) becomes $\hat{y}_k = \bar{y}_{r_i}$ for all elements $k \in s - r_i$, where $\bar{y}_{r_i} = \sum_{r_i} y_k / m_i$. This is called *respondent mean imputation*. All elements in need of imputation receive the same imputed value. The distribution of the completed data for this study variable will have a rather unnatural appearance, with a spike at $\bar{y}_{r_i}$.

Nearest neighbour imputation

In the *nearest neighbour imputation* procedure, the imputed value for element k is given by $\hat{y}_k = y_{\ell(k)}$, where $\ell(k)$ is the donor element for the nonresponding element k. That is, $\ell(k)$ provides its y-value as an imputed value for element k. The statistical idea that motivates this method is that two elements whose x-values are close should also have y-values that are close. The donor for element k is identified by distance minimization. Assuming a one-dimensional imputation variable x, define the distance from a potential donor ℓ to element k as $D_{\ell k} = |x_\ell - x_k|$. The donor $\ell(k)$ is the element belonging to the set r such that $\min_{\ell \in r} D_{\ell k}$ is obtained precisely for $\ell = \ell(k)$. That is, the distances $D_{\ell k}$ are computed for all elements $\ell \in r_i$, and the donor element for k will be the one with the minimum distance $D_{\ell k}$. For element k, we impute the donor's y-value, that is, $\hat{y}_k = y_{\ell(k)}$. Since $\ell(k)$ is closest to k, as measured by $D_{\ell k}$, it is fitting to call it the 'nearest neighbour' of k. If the imputation vector is multivariate, we can instead minimize a multivariate distance measure, for example, $D_{\ell k} = \left(\sum_{j=1}^{J} h_j (x_{j\ell} - x_{jk})^2 \right)^{1/2}$, where the quantities h_j are specified to give a suitable weighting of the J components of the vector difference $\mathbf{x}_\ell - \mathbf{x}_k$.

Hot deck imputation

In the *hot deck imputation* procedure, the imputed value for element k is $\hat{y}_k = y_{\ell(k)}$, where $\ell(k)$ is a randomly selected donor from among all potential donor elements $\ell \in r_i$. This is a donor-based, random imputation method. The distribution of the values of the resulting completed data set will look fairly natural, but may still differ considerably from the visual image obtained from the (imagined) distribution of a complete sample of y-data, $\{y_k : k \in s\}$. This is because in hot deck imputation, every donor is necessarily a respondent, and respondents and nonrespondents may differ significantly with regard to mean, variance and other characteristics.

In regression imputation and nearest neighbour imputation, the hope for close imputed values clearly rests on an assumption of a strong relationship between the study variable, y, and the imputation vector $\mathbf{x}$. Respondent mean imputation and hot deck imputation use essentially no information at all. With these weak methods, one runs a risk of imputing values that are not 'close substitutes'. Neither method is recommended when better alternatives exist. They are occasionally used as 'methods of last resort', in the absence of informative imputation variables. They will accomplish at least one of the objectives of imputation, that of creating a completed rectangular data matrix.

Imputation groups

Imputation is often performed within nonoverlapping *imputation groups*, s_g, $g = 1, \ldots, G$, whose union is the entire sample s. Within each imputation group, the imputation is carried out by one and the same method. When imputation is performed within the subgroup $s_g \subset s$, using one of the methods reviewed earlier in this section, we replace s, r_i and $s - r_i$ in the appropriate formula by, respectively, s_g, r_{ig} and $s_g - r_{ig}$, where r_{ig} is item response in group g.

We can distinguish two reasons for using more than one imputation group in the imputation process. The first reason is that different relationships are believed to exist in different subgroups of the sample. The relationship between y and the imputation vector $\mathbf{x}$ should then be formulated with this in mind. For example, if ratio imputation is used, the ratio 'sum of y_k' divided by 'sum of x_k' may differ appreciably in different subsets of the sample, suggesting a ratio imputation performed within groups. To form a relevant set of groups requires good judgement and subject-matter knowledge. Standard groupings are size by industry categories (in a business survey) or age by sex categories (in a social survey).

The second reason is the limited availability of auxiliary variables for imputation. The imputation variable(s) needed for a certain imputation method may not be available for the entire sample s. This may force a *hierarchy of imputation methods*. The stronger imputation methods are applied first, in one or more groups and for as many nonresponding elements as possible, and progressively weaker methods are applied to remaining groups. Suppose that a strongly related imputation vector **x** is available, but only for a subset of the sample elements. Then regression imputation or nearest neighbour imputation may be practised with good results for that subset. Successive groups may have to be imputed with progressively weaker **x**-vectors. More primitive methods, such as respondent mean or hot deck, may be used as a last resort for a remainder group for which little or no useful information is available.

Adding a randomly selected residual

As noted earlier, regression imputation and ratio imputation are deterministic methods: they yield the same imputed value if the process is repeated. It is of interest for some purposes to make them stochastic through the addition of a *randomly selected residual*. To illustrate, consider regression imputation and a completed data set arrived at by imputing $\hat{y}_k = \mathbf{x}'_k \hat{\boldsymbol{\beta}}_i$ given by (12.10) for those elements that require imputation. This completed data set tends to have less variability than a set of truly observed values y_k, because the regression fit that gives $\hat{y}_k = \mathbf{x}'_k \hat{\boldsymbol{\beta}}_i$ is to some degree a result of data smoothing. Adding a (randomly selected) residual will alleviate this problem. It will alter the aspect of the completed data set in the direction of a more natural variability. As a result, in the case of regression imputation, the imputed value for element k is $\hat{y}_k = \mathbf{x}'_k \hat{\boldsymbol{\beta}}_i + e_{0k}$, where $\hat{\boldsymbol{\beta}}_i = \left(\sum_{r_i} a_k \mathbf{x}_k \mathbf{x}'_k \right)^{-1} \sum_{r_i} a_k \mathbf{x}_k y_k$ as before, and e_{0k} is a randomly selected residual from the set of computed residuals $\{e_k : k \in r_i\}$, where $e_k = y_k - \mathbf{x}'_k \hat{\boldsymbol{\beta}}_i$.

The technique of adding a randomly selected residual can be practised (a) for point estimation only, (b) for variance estimation only, or (c) for both reasons. The consequence of (a) is to add variance to the imputed estimator, which may be seen as undesirable. Case (b) represents a more important use of the technique. If practised for variance estimation, an advantage is that the completed data come closer to displaying 'the natural amount of variation'. A 'standard formula' may then be adequate as part of the variance estimation procedure. We return to this issue in Section 13.4.

Remark 12.2

Imputation for qualitative variables merits a special comment. Consider the case of a dichotomous study variable that indicates the presence or the absence of a property, such as 'employed' or 'unemployed', with values 1 and 0, respectively.

To meet the requirement that the imputed value should be one that actually occurs, it must be either a 1 or a 0. An advantage of hot deck and nearest neighbour is that they satisfy this requirement. By contrast, multiple regression imputation and its special cases will usually impute values other than 0 or 1. For example, in the simple case of imputing the observed response rate within groups, $\hat{y}_k = m_g/n_g$ for all missing values in group g, we have imputed values in the interior of the unit interval. Although perhaps 'good on average', such imputation yields, for any one particular element, an 'impossible value'. The same holds when the logistic regression model is used to deliver the imputed value given for element k as $\hat{y}_k = \exp(-\mathbf{x}'_k\hat{\boldsymbol{\beta}}_i)[1 + \exp(-\mathbf{x}'_k\hat{\boldsymbol{\beta}}_i)]^{-1}$, where $\hat{\boldsymbol{\beta}}_i$ is a fitted parameter vector based on data for the elements k in the set r_i. As long as imputation is only used to produce statistics for aggregates of elements, there is no clear-cut disadvantage in imputing 'impossible values'.   □

Remark 12.3

In the technique known as *multiple imputation*, several imputations are made for the same element in the survey data set. This contrasts with the single-value imputation methods discussed so far. In multiple imputation, two or more imputations are made for a missing value. This leads to several different completed data sets. Suppose that three such sets are produced. The y_k-values for respondents are the same in all three, but the imputed values (for item nonresponse and/or for unit nonresponse) are different in the three sets. This assumes that a random imputation technique is used, for example, hot deck imputation. (Deterministic methods such as nearest neighbour and regression imputation do not qualify, because they give one and the same value in repeated attempts, unless the method is suitably modified.) The multiple imputation technique was proposed by Rubin (1978) and is discussed extensively in Rubin (1987). Multiple imputation is intended for both point estimation and variance estimation. One of the principal advantages lies in the variance estimation, which becomes very simple, given the existence of several completed data sets. In national statistical agencies, multiple imputation has so far found relatively little use. One reason may be that the method makes heavy demands on data storage space (even though only the imputed values differ from one set to the next). Multiple imputation is used in secondary analyses of survey data, that is, the kinds of analysis often carried out by analysts external to a large data producer, such as a national statistical agency.   □

12.8. IMPUTATION BY EXPERT JUDGEMENT OR HISTORICAL DATA

In statistical agencies, imputation is usually motivated by a desire to provide the 'best possible imputed value' on an *element by element basis*. The quest is for high-quality data at the micro level, rather than at some aggregated level.

Section 12.1 made an important distinction between imputation by a statistical rule and special imputation. For most elements, in particular small to medium-sized elements, the statistical rules may serve well. A practical advantage is that they can usually be carried out mechanically by existing computer software. They appeal to a perceived similarity among elements or to the kind of regularity expressed by a regression equation. One may consider members of a specific group to be similar, to the extent that the respondent mean in the group will provide adequate imputation. Or one may consider that the smoothed value produced by a linear or nonlinear regression fit is adequate imputation. But the situation is very different when one single element must be considered separately for imputation, without any apparent reference group of 'similar elements'. The element may be very large or in some other way unique. Special imputation, by expert judgement or by historical information, becomes a necessity. Such imputation is usually reserved for a small number of influential elements, and is handled by experts, paying special attention to particular features of every such element.

The following example illustrates the difference between statistical imputation and special imputation.

Example 12.2.   Statistical imputation contrasted with special imputation

Large elements are usually influential. Their impact on published survey estimates can be considerable. Consider a business survey in which a very large enterprise happens to be a nonrespondent and requires imputation. We can compare it to enterprises in the same industry group (as expressed by the Standard Industrial Classification), but a simple imputation based on the respondent mean for this group would yield a large negative imputation error, $\hat{y}_k - y_k$, for this very large element. Similarly, the respondent mean for a group of enterprises defined as 'large' in some general sense might also be misleading, because an enterprise that is 'large' in one industry may be very different (in terms of gross business income, for example) from one that is 'large' in another industry. The respondent mean of an industry-by-size group may be an unsatisfactory imputation. The imputed value from a donor identified as the 'closest neighbour' may also be inadequate, because in the upper tail of the distribution even the closest element can be numerically very different. A better approach is often a combination of historical examination and expert, but subjective, judgement. One may start by examining the series of earlier reported values, especially the most recently reported value, and adjust it in the light of the best available judgement about trends in the industry and in the economy in general. The justification is that truly large enterprises are often so unique that none of the statistical rules are likely to 'come close'. But even though the highest skill and the best judgement enter into the imputation process, we have to live with the possibility that a large error in an estimated total may be attributable to a large imputation error in one single but highly influential element.   □

CHAPTER 13

Variance Estimation in the Presence of Imputation

13.1. ISSUES IN VARIANCE ESTIMATION UNDER THE FULL IMPUTATION APPROACH

For purposes of point estimation, it is common to treat the completed data set as a set of actually observed values. One example of this is when we compute the GREG formula on the completed data set (12.1) and obtain the fully imputed GREG estimator (12.3). For variance estimation it is different. For example, when full imputation is practised, it is well known that a 'standard variance formula' may give a misleading measure of precision if applied uncritically to a completed data set. It is usually an underestimate.

The GREG estimator (4.9) for complete response is accompanied by a corresponding variance estimator formula. This 'standard formula' is appropriate for computing variance estimates and confidence intervals, for complete response, according to the recipe

$$\text{point estimate} \pm 1.96 \, (\text{variance estimate})^{1/2}.$$

In repeated samples drawn by the given sampling design, this interval will cover the parameter value for roughly 95 % of all samples. It is a *design-based confidence interval*, because the coverage rate refers to repeated samples drawn with the given design. The roughly 95 % coverage rate is achieved for any shape of the sampled finite population (with the possible exception of highly abnormal shapes).

Nonresponse brings additional variance over and above the sampling variance. When imputation is used, this is not always recognized. Some users may argue that the variance increase is in any case small and can be ignored. This may be true for a modest imputation rate such as 3 %, but not when the rate is 30 %, as it is in many surveys.

It is not unusual, in many statistical agencies, to treat imputed values as observed values also for purposes of variance calculation. That is, the variance formula for complete response is taken as a *prototype* and is computed on the n values of the completed data set (12.1), in the optimistic but often misguided

Estimation in Surveys with Nonresponse   C.-E. Särndal and S. Lundström
© 2005 John Wiley & Sons, Ltd

belief that this will give a sufficiently good indication of the variance of the imputed estimator. This approach of 'acting as if the imputed values are perfect substitutes' leads to unsatisfactory variance estimates, for two reasons:

(i)   The prototype formula, computed on the completed data set, gives a biased estimate of the sampling variance, the bias being usually negative.

(ii) No attempt is made to account for the additional component of variance caused by the nonresponse.

Consequently, the computed confidence intervals will be misleading and the bias in the point estimator is not the only reason. On average, the intervals are too short. The interval based on the normal score 1.96 will *not* cover the parameter value in roughly 95 % of all cases, as is the intent. The coverage rate is less than 95 %.

That the prototype formula may give a misleading indication of the variance of the imputed estimator is illustrated by the following simple example.

Example 13.1.   'Right amount of variation' in the completed data?

For simplicity, consider SI with the sampling fraction n/N. Suppose first that the survey has complete response and that we estimate the population total Y by $\hat{Y}_{HT} = N\bar{y}_s$, where $\bar{y}_s = \sum_s y_k/n$ is the sample mean. The well-known expression for the design-based variance is $V(\hat{Y}_{HT}) = N^2(1/n - 1/N)S_{yU}^2$, where S_{yU}^2 is the population variance of the N values y_k. The unbiased estimator of the variance, computed on the sample, is $\hat{V}(\hat{Y}_{HT}) = N^2(1/n - 1/N)S_{ys}^2$, where S_{ys}^2 is the variance of the n observed values y_k.

Now turn to the situation with nonresponse and imputation (by any suitable method). The complete response estimator $\hat{Y}_{HT} = N\bar{y}_s$ leads to the fully imputed estimator $\hat{Y}_{IHT} = N\bar{y}_{\bullet s}$, where $\bar{y}_{\bullet s} = \sum_s y_{\bullet k}/n$ is the mean of the completed data set defined by (12.1). When computed on the completed data set, the prototype formula $N^2(1/n - 1/N)S_{ys}^2$ gives $N^2(1/n - 1/N)S_{y\bullet s}^2$, where $S_{y\bullet s}^2 = \sum_s (y_{\bullet k} - \bar{y}_{\bullet s})^2/(n-1)$ is the variance of the n values $y_{\bullet k}$ in the completed data set. What can be said about $S_{y\bullet s}^2$? It depends on the imputation method whether or not $S_{y\bullet s}^2$ is close to the desired value S_{ys}^2. The imputed values may not contain sufficient variation for this to happen. A case in point is respondent mean imputation, which is not recommended except as a last resort. In that case $\hat{y}_k = \bar{y}_{r_i} = \sum_{r_i} y_k/m_i$ for every $k \in s - r_i$, where r_i is the item response set and m_i its size. The fully imputed estimator becomes $\hat{Y}_{IHT} = N\bar{y}_{r_i}$. Computing the prototype formula $N^2(1/n - 1/N)S_{ys}^2$ on the completed data set gives

$$N^2 \left(\frac{1}{n} - \frac{1}{N}\right) S_{y\bullet s}^2 = N^2 \left(\frac{1}{n} - \frac{1}{N}\right) \frac{m_i - 1}{n - 1} S_{yr_i}^2 \qquad (13.1)$$

where $S_{yr_i}^2 = \sum_{r_i}(y_k - \bar{y}_{r_i})^2/(m_i - 1)$.

An immediate reaction to this result is that is looks too small as an indicator of the sampling variance of $N\bar{y}_{r_i}$. More reasonable would be $N^2(1/n - 1/N)S_{yr_i}^2$. This would in fact be perfect if nonresponse occurs completely at random,

because then $S^2_{yr_i}$ equals S^2_{ys} on average. Thus with respondent mean imputa-
tion, $N^2(1/n - 1/N)S^2_{y \bullet s}$ will understate the more reasonable value $N^2(1/n - 1/N)S^2_{yr_i}$ by a factor of $(m-1)/(n-1) \approx m_i/n$ If the response rate m_i/n is
70 %, the understatement is as much as 30 %. But in addition, $\hat{Y}_{\text{IHT}} = N\bar{y}_{r_i}$ has
a second component of variance, due to nonresponse. This component must also
be covered in order to account for the whole variance of $N\bar{y}_{r_i}$. We could accept

$$N^2 \left(\frac{1}{m_i} - \frac{1}{N} \right) S^2_{yr_i} \tag{13.2}$$

as an estimate of the *total* variance, based on the reasoning that the full imputation
reduces the effective sample size from n to m_i, and (13.2) is reminiscent of
'm_i elements drawn from N'. Thus the result (13.1) obtained by the prototype
formula is smaller than the reasonable estimate (13.2) by a factor of roughly
$(m_i/n)^2$, which is only 49 % when the response rate is 70 %. □

Although the conditions are extremely simple in Example 13.1, it reveals a risk
of variance underestimation, unless proper attention is paid. We shall examine
the combined approach, and in particular the question of variance estimation for
that approach.

13.2. AN IDENTITY OF COMBINED AND FULLY WEIGHTED APPROACHES

The combined approach and the fully weighted approaches lead, in general, to
different estimators, $\hat{Y}_{\text{IW}}$ and $\hat{Y}_{\text{FW}}$, given, respectively, by (12.7) and (12.9). The
former may be preferred for practical reasons, as mentioned. However, they
are identical under an appropriately specified regression imputation. Suppose we
use regression imputation for the item nonresponse as follows: for $k \in r - r_i$,
impute

$$\hat{y}_k = \mathbf{x}'_k \hat{\boldsymbol{\beta}}_i \tag{13.3}$$

with

$$\hat{\boldsymbol{\beta}}_i = \left(\sum_{r_i} d_k \mathbf{z}_k \mathbf{x}'_k \right)^{-1} \sum_{r_i} d_k \mathbf{z}_k y_k. \tag{13.4}$$

This agrees with the regression imputation (12.10) if we set $a_k \mathbf{x}_k = d_k \mathbf{z}_k$. The
standard is $\mathbf{z}_k = \mathbf{x}_k$, and otherwise $\mathbf{z}_k$ is the same instrument vector as in (12.4)
and (12.8). The imputation is study variable specific, because r_i and $\hat{\boldsymbol{\beta}}_i$ change
from one y-variable to the next. An identity of combined and fully weighted
approaches is now expressed in the following proposition.

Proposition 13.1

Consider (i) the item imputed calibration estimator (12.7),

$$\hat{Y}_{\text{IW}} = \sum_r d_k v_k y_{\bullet k} = \sum_{r_i} d_k v_k y_k + \sum_{r-r_i} d_k v_k \hat{y}_k$$

with the weight system $d_k v_k$ given for $k \in r$ by (12.4), and (ii) the fully weighted calibration estimator (12.9),

$$\hat{Y}_{\text{FW}} = \sum_{r_i} d_k v_{ik} y_k$$

with the weight system $d_k v_{ik}$ given for $k \in r_i$ by (12.8). Then, under the regression imputation defined by (13.3) and (13.4), $\hat{Y}_{\text{IW}}$ and $\hat{Y}_{\text{FW}}$ are identical, that is, $\hat{Y}_{\text{IW}} = \hat{Y}_{\text{FW}}$ for every outcome s, r and r_i.   □

The proof is given in the end-of-chapter appendix.

Under conditions somewhat different from ours, Gabler and Häder (1999) establish a similar equivalence between an imputed GREG estimator and a calibration estimator.

Proposition 13.1 is important because it opens up a technique for estimating the variance of the item imputed estimator $\hat{Y}_{\text{IW}}$. The identity of $\hat{Y}_{\text{IW}}$ and the fully weighted estimator $\hat{Y}_{\text{FW}}$ allows a use of Proposition 11.1, which addresses a fully weighted estimator. The procedure is as follows.

Procedure 13.1

For the combined approach, suppose that we treat the item nonresponse by the regression imputation (13.3) and (13.4) and then compute the item imputed estimator $\hat{Y}_{\text{IW}}$. By Proposition 13.1, $\hat{Y}_{\text{IW}} = \hat{Y}_{\text{FW}}$ for any outcome s, r and r_i, where v_{ik} is given by (12.8). But $\hat{Y}_{\text{FW}}$ is a calibration estimator, so the variance estimator formula (11.3) in Proposition 11.1 applies to $\hat{Y}_{\text{IW}}$ with one simple modification: substitute r_i for r everywhere in (11.1) and (11.3)–(11.6). As a result, the residuals $\hat{e}_k^*$ and $\hat{e}_k$ in these formulae will depend on the specific y-variable under consideration.   □

Variance estimation by Procedure 13.1 requires a different computation set r_i for each study variable. The procedure points out m_i, the size of r_i, as the 'effective sample size' of $\hat{Y}_{\text{IW}} = \hat{Y}_{\text{FW}}$, rather than the greater size m of r, notwithstanding the fact that $\hat{Y}_{\text{IW}}$ is computed as a sum over the whole response set r. If the study variable y were free of item nonresponse, we would be in a position to use $\hat{Y}_{\text{W}}$ given by (12.5), and then the effective size would indeed be m. But imputing for $m - m_i$ missing values will, so to speak, cause a loss of equally many effective observations, or degrees of freedom, compared to when the variable is free of item nonresponse. Item nonresponse causes the effective response set to 'shrink' from r to r_i. The resulting increase in variance must be reflected in the variance estimation procedure.

Remark 13.1

Estimation for a domain requires special attention. The identity in Procedure 13.1 applies to any variable y, thus it applies in particular to the domain-specific variable y_q, whose value is defined as $y_{qk} = \delta_{qk} y_k$, where $\delta_{qk} = 1$ for k inside the domain U_q and $\delta_{qk} = 0$ outside. The identity of $\hat{Y}_{q\text{IW}}$ and $\hat{Y}_{q\text{FW}}$ is realized if we define the instrument vector in (13.4) as $\mathbf{z}_k = \delta_{qk}\mathbf{x}_k$. This amounts to a separate regression imputation for each domain.   □

Remark 13.2

Both $\hat{Y}_{\text{IW}} = \hat{Y}_{\text{FW}}$ (based on an effective sample size of m_i) and $\hat{Y}_{\text{W}}$ (based on a greater effective sample size of m) have the bias given in Proposition 9.1 as $-\sum_U (1 - \theta_k) e_{\theta k}$ with $e_{\theta k} = y_k - \mathbf{x}'_k \mathbf{B}_{\text{U};\theta}$, where $\mathbf{B}_{\text{U};\theta} = \left(\sum_U \theta_k \mathbf{z}_k \mathbf{x}'_k \right)^{-1}$ $\sum_U \theta_k \mathbf{z}_k y_k$. We thus conclude that, as long as the set r_i is not too small, the bias stays the same whether there is item nonresponse or not, assuming that the regression imputation (13.3) is used. The variance does increase with increasing item nonresponse, because the effective sample size is reduced.   □

Example 13.2

Let us carry out Procedure 13.1 under the simple conditions of SI sampling of n from N, and $\mathbf{x}_k = \mathbf{x}^*_k = \mathbf{z}_k = 1$ for all k. This is a weak auxiliary vector; the imputed values are not of high quality, but the results are easy to interpret. For the combined approach, (12.4) gives $v_k = n/m$, hence $w_k = d_k v_k = (N/n)(n/m) = N/m$ for all k, and (13.3) and (13.4) lead to imputation by the respondent mean, $\hat{y}_k = \bar{y}_{r_i} = \sum_{r_i} y_k/m_i$, for every $k \in r - r_i$. The item imputed calibration estimator is thus $\hat{Y}_{\text{IW}} = \hat{Y}_{\text{FW}} = N\bar{y}_{r_i} = (N/m_i) \sum_{r_i} y_k$. Denote by $S^2_{y r_i} = (m_i - 1)^{-1} \sum_{r_i} (y_k - \bar{y}_{r_i})^2$ the variance of the m_i effective observations. The estimated variance components obtained by Procedure 13.1 are

$$\hat{V}_{\text{SAM}} = N^2 \left(\frac{1}{n} - \frac{1}{N} \right) Q_{r_i} S^2_{y r_i} \approx N^2 \left(\frac{1}{n} - \frac{1}{N} \right) S^2_{y r_i}$$

and

$$\hat{V}_{\text{NR}} = N^2 \left(\frac{1}{m_i} - \frac{1}{n} \right) R_{r_i} S^2_{y r_i} \approx N^2 \left(\frac{1}{m_i} - \frac{1}{n} \right) S^2_{y r_i}$$

where

$$Q_{r_i} = \frac{m_i - 1}{m_i} \frac{n}{n - 1} \left(1 - \frac{1}{n} + \frac{1}{m_i} \right) \approx 1, \qquad R_{r_i} = \frac{m_i - 1}{m_i} \approx 1.$$

Thus

$$\hat{V}(N\bar{y}_{r_i}) = \hat{V}_{\text{SAM}} + \hat{V}_{\text{NR}} \approx N^2 \left(\frac{1}{m_i} - \frac{1}{N} \right) S^2_{y r_i}.$$

This is a credible expression for the estimated variance of $N\bar{y}_{r_i}$, considering that only m_i (and neither m nor n) out of the N values y_k have been observed. If the nonresponse occurs completely at random, the result is without reproach. In practice, we would seek a better auxiliary vector than the primitive $\mathbf{x}_k = 1$, and the accuracy of the variance estimation would improve.   □

Example 13.3

Another important sampling design that gives transparent results is STSI. Consider the auxiliary vector $\mathbf{x}_k = \mathbf{x}^*_k = \mathbf{z}_k = \gamma_k$, where γ_k is the one-way classification vector that indicates the strata. That is, each stratum coincides with a group coded by the classification vector γ_k. Denote by n_h/N_h the sampling rate and by r_h the response set of size m_h in stratum h. The effective sample size in stratum h is m_{ih}, the size of $r_{ih} = r_h \cap r_i$. This is the set of elements in stratum h for which we have observed values for the study variable y. In this case, (13.3) and (13.4) give the imputation $\hat{y}_k = \bar{y}_{r_{ih}}$ for every $k \in r_h - r_{ih}$, where $\bar{y}_{r_{ih}} = \sum_{r_{ih}} y_k/m_{ih}$ is the mean of the m_{ih} values observed in stratum h. From (12.4) the weights are $d_k v_k = N_h/m_h$ for all k in stratum h. As we can expect, the item imputed calibration estimator (12.7) becomes a weighted sum over the H strata: $\hat{Y}_{IW} = \hat{Y}_{FW} = \sum_{h=1}^{H} N_h \bar{y}_{r_{ih}}$. Let $S^2_{yr_{ih}}$ be the variance of the m_{ih} observed y-values in stratum h. Procedure 13.1 leads after some algebra to

$$\hat{V}_{SAM} \approx \sum_{h=1}^{H} N_h^2 \left(\frac{1}{n_h} - \frac{1}{N_h} \right) S^2_{yr_{ih}}$$

$$\hat{V}_{NR} \approx \sum_{h=1}^{H} N_h^2 \left(\frac{1}{m_{ih}} - \frac{1}{n_h} \right) S^2_{yr_{ih}}.$$

Adding the two components gives

$$\hat{V} \left(\sum_{h=1}^{H} N_h \bar{y}_{r_{ih}} \right) \approx \sum_{h=1}^{H} N_h^2 \left(\frac{1}{m_{ih}} - \frac{1}{N_h} \right) S^2_{yr_{ih}}.$$

The result is easy to interpret. We can associate it with a simple random selection of m_{ih} from the N_h elements within stratum h.   □

13.3. MORE ON THE RISK OF UNDERESTIMATING THE VARIANCE

Example 13.1 hinted that imputation meets with a certain risk of underestimating the variance, unless proper attention is paid. Procedure 13.1 and Example 13.2 show that variance estimation can be carried out properly, at least under regression imputation. The following example throws further light on the question of understating the variance.

Example 13.4

Consider again the simple conditions of Example 13.2 with SI sampling of n elements from N, and $\mathbf{x}_k = \mathbf{x}_k^* = \mathbf{z}_k = 1$ for all k. The imputed values are $\hat{y}_k = \bar{y}_{r_i}$ for $k \in r - r_i$, where r is the survey response set of size m and r_i is the item-specific response set of size m_i, where $m_i < m$. The combined approach gives the item-imputed estimator $\hat{Y}_{IW} = N\bar{y}_{r_i}$. The corresponding variance estimate, obtained by Procedure 13.1 and derived in Example 13.2, is

$$\hat{V}(N\bar{y}_{r_i}) \approx N^2 \left(\frac{1}{m_i} - \frac{1}{N} \right) S^2_{yr_i}. \qquad (13.5)$$

We can accept (13.5) as a reasonable variance estimator. It embodies the idea of a simple random selection of m_i y-values out of N.

Alternative approaches may come to mind for to estimating the variance of $N\bar{y}_{r_i}$. Let us examine the idea of computing variance by a prototype formula. The point estimator $\hat{Y}_W$ given by (12.5) addresses a situation characterized by some unit nonresponse but an absence of item nonresponse. If these conditions apply, an acceptable variance estimator, mentioned in Example 13.1, would be $N^2\,(1/m - 1/N)\,S^2_{yr}$, where m is the size of the survey response set r. Let us take this formula as a prototype for variance calculation for $\hat{Y}_{IW} = N\bar{y}_{r_i}$. We compute the formula on the data completed for item nonresponse, $y_{\bullet k} = y_k$ for $k \in r_i$ and $y_{\bullet k} = \bar{y}_{r_i}$ for $k \in r - r_i$. Then S^2_{yr} is replaced by

$$S^2_{y\bullet r} = \frac{1}{m-1}\sum_r (y_{\bullet k} - \bar{y}_{\bullet r})^2 = \frac{m_i-1}{m-1}S^2_{yr_i}.$$

The resulting variance estimator for $\hat{Y}_{IW} = N\bar{y}_{r_i}$ is

$$N^2\left(\frac{1}{m} - \frac{1}{N}\right)\frac{m_i-1}{m-1}S^2_{yr_i}. \tag{13.6}$$

Compared with the more adequate estimate (13.5), formula (13.6) is an understatement by a factor of roughly $(m_i/m)^2$. The understatement is not severe if the item nonresponse rate is modest.

Yet another approach, still less satisfactory, is to fetch the prototype variance formula from the case of an SI selection with a complete response. This formula is $N^2(1/n - 1/N)S^2_{ys}$, and suppose we compute it on the data set completed for *both* unit *and* item nonresponse, that is, $y_{\bullet k} = y_k$ for $k \in r_i$ and $y_{\bullet k} = \bar{y}_{r_i}$ for $k \in s - r_i$. The result, mentioned in Example 13.1, is

$$N^2\left(\frac{1}{n} - \frac{1}{N}\right)\frac{m_i-1}{n-1}S^2_{yr_i}. \tag{13.7}$$

This understates the more adequate estimate (13.5) by a factor of roughly $(m_i/n)^2$, a stronger understatement than (13.6).

To illustrate, suppose the unit nonresponse rate is $(n - m)/n = 22\%$ and the item nonresponse rate $(m - m_i)/n = 8\%$. The total nonresponse rate is 30%. Then the variance estimate (13.6) is roughly $(0.70/0.78)^2 = 80.5\%$ of the more appropriate estimate (13.5). By contrast, (13.7) is more severely handicapped. It is only about $(0.70)^2 = 49\%$ of estimate (13.5).

Variance estimation by a prototype formula amounts to treating the completed data set as a satisfactory substitute for a set consisting entirely of observed data. Some caution is needed. There is a risk of understating the variance. In particular, the prototype that accompanies estimator (12.5), which addresses 'unit nonresponse but no item nonresponse', may deliver some understatement of the variance of the combined approach estimator (12.7). The understatement is minor for modest item nonresponse rates. □

13.4. A BROADER VIEW OF VARIANCE ESTIMATION FOR THE COMBINED APPROACH

Variance estimation for the combined approach estimator $\hat{Y}_{IW}$ was developed in Procedure 13.1 under the particular conditions of regression imputation. For other methods than regression imputation, how should we estimate the variance of $\hat{Y}_{IW}$? This important question is complicated by the fact that more than one imputation method is often used in practice to produce the imputed values $\hat{y}_k$ for $k \in r - r_i$. Some of the values may be regression imputed, others nearest neighbour imputed, still others imputed by expert judgement, and so on.

A definitive answer is not available. But there exist several computational openings. We outline three options, assuming that the information input for $\hat{Y}_{IW}$ is $\mathbf{X} = \begin{pmatrix} \mathbf{X}^* \\ \hat{\mathbf{X}}^\circ \end{pmatrix}$. From a theory standpoint, all are 'approximate procedures'.

(i) Relying on Procedure 13.1. Although developed with regression imputation in mind, we can execute the procedure even though only some, or none at all, of the item nonresponse is regression imputed. There are no computational obstacles. The two variance components can be computed for any item response set r_i and for any set of data completed for item nonresponse, $\{y_{\bullet k} : k \in r\}$, where $y_{\bullet k}$ is the observed value y_k for $k \in r_i$ and an imputed value $\hat{y}_k$ for $k \in r - r_i$. The inherent assumption is that, regardless of the imputation method(s), the set of imputed values $\hat{y}_k$ for $k \in r - r_i$ is roughly substitutable for a set of regression imputed values. The procedure does recognize the reduction of the effective response set from r to the smaller set r_i. This is important to counteract the tendency to understatement of variance.

(ii) Use of a prototype variance estimator. The estimator $\hat{Y}_W$ given by (12.5) is appropriate when the survey may have unit nonresponse but the variable y is free of item nonresponse, so that $r = r_i$. The variance estimator is given by (11.3) in Proposition 11.1. We can take (11.3) as a prototype and compute it on the data completed for item nonresponse, $\{y_{\bullet k} : k \in r\}$. This option may cause some underestimation of variance of $\hat{Y}_{IW}$. However, if the item nonresponse rate is modest, the underestimation is limited and perhaps inconsequential.

(iii) Use of a prototype formula, following an adjustment of the completed data set. The understatement of variance in option (ii) may be traceable to a lack of 'sufficient variation' in the imputed values $\hat{y}_k$ for $k \in r - r_i$. We can then compute a prototype formula as in option (ii), but after having introduced 'sufficient variation' into the imputed values, for purposes of variance computation only. This counteracts the tendency to understatement. If the $\hat{y}_k$ are regression imputed values, the technique for adding a randomly selected residual was explained in Section 12.7. With such adjustment of the completed data set, option (iii) is less susceptible to underestimation than (ii).

A positive aspect of all three options is that they are simple to carry out with existing software, CLAN97 and others. Although all three rely, to different degrees, on approximation, they will provide a valuable indication of the variance, albeit not an unbiased estimation of the variance. It should be borne in mind that the accuracy, measured by the mean squared error, is the sum of variance and squared bias. The squared bias, in many cases the larger component of the two, cannot in any case be measured unbiasedly. In light of this, the objective, although justified, of a precise measurement of variance loses some of its urgency.

13.5. OTHER ISSUES ARISING WITH REGARD TO ITEM NONRESPONSE

Most surveys are affected by both item nonresponse and unit nonresponse. For the treatment of the item nonresponse, we have considered two alternatives, the combined approach, in Section 12.5, and the fully weighted approach, in Section 12.6. Of the two, the combined approach is often favoured. It uses a rectangular data matrix. Every study variable in the survey will have values recorded (by observation or by imputation) for every element k in the response set r. A unique set of calibrated weights can be applied to all study variables. A few additional points need to be addressed in regard to the treatment of item nonresponse.

Discarding information on item nonresponse elements

One approach sometimes employed simply ignores study variable values observed for the item nonresponse elements. No imputation occurs. In this case, too, the result is a rectangular data set. This voluntary sacrifice of data may cause a substantial loss of information. The technique is not recommended unless the item nonresponse set is very small. An argument can be made that all data in the remaining rectangular matrix are really observed data. Therefore, one avoids the distortion of the distribution of a y-variable, or the relationship between two or more y-variables, that accompanies some of the imputation procedures.

The 'data missing' category

For a categorical variable, a survey questionnaire will indicate a set of exhaustive categories or response alternatives. A respondent indicates one of these 'real categories'. If the sampled person k indicates none, he or she is registered as nonrespondent on that categorical variable. A technique sometimes practised in such circumstances is to create a special 'data missing' category. It coexists with the real categories and is treated in the estimation process as any real category. The population counts for the real categories tend to be underestimated. For parameters defined as proportions (ratios of counts), the estimate is constructed as the ratio of the 'weighted estimate of the number of elements in the category' to the 'weighted estimate of the total number of elements.' Although both the numerator and the denominator may be underestimates, the ratio may be little affected. The principal motivation seems to be to avoid the use of imputation; the procedure is not recommended.

13.6. FURTHER COMMENTS ON IMPUTATION

The specialized literature on imputation contains a more in-depth discussion of the important issues mentioned in this section.

Distortion of the distribution of a variable

Some imputation methods are prone to create a more or less pronounced distortion of the distribution of a study variable. This is strikingly obvious if the overall respondent mean is imputed. A perhaps considerable number of elements will be assigned one and the same value, namely, the respondent mean. The distortion is less pronounced, but still noticeable, when imputation is instead by the respondent mean within groups, or by multiple regression imputation. Even a multiple regression imputation based on several x-variables tends to deliver a completed data set with unnaturally low variability, as compared to a data set resulting from complete response. Of the methods discussed in Chapter 12, nearest neighbour imputation seems to be the least susceptible to this drawback. It comes closest to rendering a natural distribution and a natural variability in the completed data set. Whatever the imputation method used, the completed data set should be subjected to the usual checks for internal consistency. All imputed values should undergo the editing checks normally carried out for the survey.

Distortion of the relationship among variables

The relationships between variables are also likely to be more or less perturbed by imputation. Consequently, regression analysis or other types of analysis can give misleading results, compared to an analysis based on complete response on all variables involved. Nearest neighbour imputation is likely to be the method offering the best protection against this disadvantage.

Estimation for domains

In estimation for domains, a contrast emerges between weighting on the one hand and imputation on the other. It is often desirable to impute for the item nonresponse, rather than to use full weighting. Both the bias and the variance of an estimator can be lower with imputation. The reason becomes apparent if we contrast the idea of imputation with that of weighting. For example, for a large and influential element, imputation by expert judgement may yield a close (albeit not perfect) surrogate value. It stands a better chance of reflecting the unique character of the element than the expansion of observed study variable values through calibrated weights. The calibrated weights always involve a degree of smoothing, with reference to a set of other elements.

APPENDIX: PROOF OF PROPOSITION 13.1

Let us develop each of the two sums on the right-hand side of $\hat{Y}_{\text{IW}} = \sum_{r_i} d_k v_k y_k + \sum_{r-r_i} d_k v_k \hat{y}_k$ given by (12.7). In the first sum, insert the expression (12.4) for v_k to obtain

$$\sum_{r_i} d_k v_k y_k = \sum_{r_i} d_k y_k + G \qquad (A13.1)$$

where $G = \left(\mathbf{X} - \sum_r d_k \mathbf{x}_k \right)' \left(\sum_r d_k \mathbf{z}_k \mathbf{x}_k' \right)^{-1} \sum_{r_i} d_k \mathbf{z}_k y_k$. The second sum is $\sum_{r-r_i} d_k v_k \hat{y}_k = \sum_r d_k v_k \mathbf{x}_k' \hat{\boldsymbol{\beta}}_i - \sum_{r_i} d_k v_k \mathbf{x}_k' \hat{\boldsymbol{\beta}}_i$. Transform the terms on the right-hand side by inserting the expression (12.4) for v_k, and use two facts: (a) that

$\sum_r d_k v_k \mathbf{x}_k = \mathbf{X}$ by calibration; and (b) that $\sum_{r_i} d_k \mathbf{z}_k \mathbf{x}'_k \hat{\boldsymbol{\beta}}_i = \sum_{r_i} d_k \mathbf{z}_k y_k$. This leads to

$$\sum_{r-r_i} d_k v_k \hat{y}_k = \left(\mathbf{X} - \sum_{r_i} d_k \mathbf{x}_k \right)' \hat{\boldsymbol{\beta}}_i - G. \tag{A13.2}$$

Summing (A13.1) and (A13.2) gives $\hat{Y}_{\text{IW}} = \sum_{r_i} d_k y_k + (\mathbf{X} - \sum_{r_i} d_k \mathbf{x}_k)' \hat{\boldsymbol{\beta}}_i$. But considering the definition (12.8) of v_{ik}, we can equivalently write $\hat{Y}_{\text{IW}} = \sum_{r_i} d_k v_{ik} y_k$, and therefore $\hat{Y}_{\text{IW}} = \hat{Y}_{\text{FW}}$ for any outcome s, r and r_i, as claimed in the proposition.

CHAPTER 14

Estimation Under Nonresponse and Frame Imperfections

14.1. INTRODUCTION

We have noted in the preceding chapters that both the sampling error and the nonresponse error can be substantially reduced when powerful auxiliary information is available and used in a calibration weighting approach. In this chapter, we face the additional problem of coverage error.

Some important terminology and notation with regard to frame imperfections was introduced in Section 2.2. We use that terminology here. Many surveys are affected by frame imperfections, often referred to as coverage errors. The frame can have undercoverage, overcoverage or both. Thus the estimation procedure needs to deal simultaneously and effectively with three types of error: sampling error, nonresponse error and coverage error. In previous chapters we have considered the first two of these.

It is not a trivial step to accommodate the third kind of error within the general framework set by the preceding chapters. The literature to date does not offer any firm guidelines. Although imperfect frames are a reality in most large surveys, the remedies employed in statistical agencies and elsewhere vary considerably. In the absence of 'firmly established methodology', the procedures in use may seem *ad hoc*.

The objective in this chapter is to extend the theory for weighting and imputation presented in earlier chapters. There are few 'conventional methods'. However, we cite a few examples of estimation procedures that are in use. They are shown to be special cases of the general approach in this chapter, which is a step towards a structured outlook.

The treatment of coverage imperfections usually requires assumptions. Their validity may be hard or impossible to verify.

As in earlier chapters we wish to estimate the *target population* total

$$Y_U = \sum_U y_k \qquad (14.1)$$

where y_k is the value of the study variable, y, for the kth element of the target population $U = \{1, \ldots, k, \ldots, N\}$. It may differ from the *frame population*

Estimation in Surveys with Nonresponse   C.-E. Särndal and S. Lundström
© 2005 John Wiley & Sons, Ltd

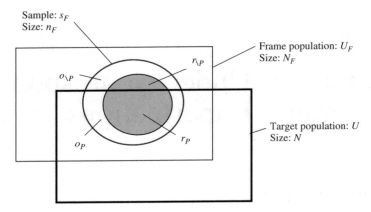

Figure 14.1   Representation of selected sample and response set, seen as subsets of the frame population, and their intersections with the target population.

from which sample selection is carried out. We address the situation shown in Figure 14.1.

The notation for totals in this chapter follows the rule that if A is a set of elements, then Y_A denotes the total $\sum_{k \in A} y_k$.

Let s_F be a sample of size n_F drawn from the frame population U_F (of size N_F) with the probability $p(s_F)$. The inclusion probabilities, known for all k, $\ell \in U_F$, are then $\pi_k = \sum_{s_F \ni k} p(s_F)$ and $\pi_{k\ell} = \sum_{s_F \supset \{k,\ell\}} p(s_F)$. Let $d_k = 1/\pi_k$ denote the *design weight* of element k and let $d_{k\ell} = 1/\pi_{k\ell}$.

As discussed in Section 2.2, the reference time for U_F (the point in time at which U_F is finalized) may precede by six months or more the reference time for U (the point in time which the survey sets out to address and measure).

The frame U_F may consist of a register that is updated more or less regularly. For example, the total population registers in the Scandinavian countries are frequently updated. They will, at most points in time, give very good coverage of the target population U. Many other surveys do not enjoy similarly favourable conditions. The coverage imperfections may be considerable.

The frame population U_F and the target population U are not identical, but they have a common part, which we denote $U_P = U \cap U_F$. The elements in U_P are 'persisters' in that they are both in the frame and in scope for the survey.

The frame population U_F in Figure 14.1, has both undercoverage and overcoverage. The *overcoverage set* is $U_F - U_P$; the *undercoverage set* is $U - U_P$. The set of elements that respond *and* belong to the target population is denoted by r_P. It is the set of responding persisters; its size is denoted m_P. We have $r_P \subseteq s_F$. We denote by $r_{\backslash P}$ the set of elements that respond *and* belong to the overcoverage. Let $m_{\backslash P}$ be the size of $r_{\backslash P}$. The subscript $\backslash P$ is to be interpreted as 'nonpersisters'.

The nonresponse set is the set of frame elements that do not respond, $o_F = o_P \cup o_{\backslash P}$, where o_P is the part that belongs to the target population and $o_{\backslash P}$ the part that belongs to the overcoverage. The sample s_F is the union of the four nonoverlapping sets $r_P, r_{\setminus P}, o_P$ and $o_{\setminus P}$. We denote the set of frame respondents (those who are in the frame and respond) by $r_F = r_P \cup r_{\setminus P}$.

We assume that every element, $k \in r_F$, can be identified as belonging to either r_P or $r_{\setminus P}$, as the case may be. This identification is usually straightforward in practice. Much more problematic in practice is the split of the set of nonresponding frame elements, o_F, into its two subsets, o_P and $o_{\setminus P}$. A reason for failure to respond may be that the element has ceased to exist in the interval between the reference time for U_F and the reference time for U. It is then an overcoverage element, but chances are that this cannot be established.

The set of observed y-data is $\{y_k : k \in r_P\}$; y_k is missing for the elements $k \in o_P$, and y-data for $k \in r_{\setminus P}$, if observed, are not used.

We consider an auxiliary vector $\mathbf{x}_k = \mathbf{x}_k^*$, whose value is known for every $k \in r_F = r_P \cup r_{\setminus P}$, and such that the frame population total $\mathbf{X}_{U_F} = \sum_{U_F} \mathbf{x}_k^*$ is also known. It can be obtained by a simple addition if $\mathbf{x}_k^*$ is specified in the frame U_F for every k. More generally, additional auxiliary information may exist in the form of known values of a vector $\mathbf{x}_k^\circ$ for every $k \in s_F$, but for simplicity we do not consider this possibility here.

The estimation of the target population parameter $Y_U = \sum_U y_k$ faces the following problems:

(i)   the absence of observed y-data from the undercoverage set;
(ii) the absence of a correct auxiliary vector total for the target population;
(iii) the difficulty of decomposing the nonresponse set o_F into its subsets o_P and $o_{\setminus P}$.

We shall examine two different procedures for estimating the target population total Y_U. Neither solution is perfect, but both have a background in actual practice. Both are hampered, in different ways, by problems (i), (ii) and (iii).

1   Estimation of Y_U by the sum of two terms, as detailed in Section 14.2. The two terms are: (1a) an estimate of the persister total $Y_{U_P} = \sum_{U_P} y_k$; and (1b) a separate term to compensate for the undercoverage total $Y_{U - U_P}$. In Section 14.2 we assume that the compensation in (1b) has been carried out, and we address the simpler problem of estimating Y_{U_P}.
2   Direct estimation of Y_U, considered in Section 14.3. Here, the construction of the estimator aims directly at the target population total Y_U, despite the obvious difficulty posed, in particular, by problem (i).

The choice between procedures 1 and 2 depends on the information available in the survey. The compensation in step (1b) is problematic. In some surveys, it may be possible to carry out a supplementary sample selection from the undercoverage set and to base the compensation term on the information thus gathered. Failing this, one must rely on assumptions and sound professional judgement. Information may be used both from the current occasion and from earlier survey occasions. One can expect some bias in the estimates. A theoretically flawless approach to step (1b) is presently lacking. If a satisfactory compensation can be obtained for Y_{U-U_P}, it is reasonable to choose procedure 1. In other cases, the auxiliary information at hand may favour the use of procedure 2, carried out with a calibration approach, as Section 14.3 explains.

Problem (iii) has particularly grave consequences when imputation forms part of the whole estimation effort. Imputation for nonresponse elements makes good sense only if we can correctly identify the set o_P, that is, those nonresponding elements which are also in the target population. As noted in problem (iii), this may not be possible in practice. Problem (iii) also affects the variance estimation for the calibration approach, as will be discussed in Section 14.3.

In countries endowed with a total population register, many surveys on individuals and households rely on sampling from that register. As discussed in Example 2.1, such a register is updated at regular intervals, perhaps as frequently as once a month. Consequently, in a matching of the sample s_F with a very recently updated version of the register, the set o_P of nonrespondents in scope for the survey will be identified with little error. Problem (iii) is then less onerous. Even though a couple of weeks may have elapsed since the last update, the coverage errors (the difference between the updated register and the target population) may be considered inconsequential.

In enterprise surveys, the situation is often less favourable. A business register, discussed in Example 2.2.2, is usually updated much less frequently, perhaps only once a year, and even after such an updating, the register is not flawless. At most points in a calendar year a highly accurate identification of o_P may not be possible.

We assume in this chapter that item nonresponse has been treated in a preliminary step, by imputation. The y-data are assumed to consist of a majority of observed values plus some imputed values for the item nonresponse. Weighting by calibration compensates for the unit nonresponse.

14.2. ESTIMATION OF THE PERSISTER TOTAL

The persister set $U_P = U \cap U_F$ defines a domain of the target population U. It is also a domain of the frame population U_F. When the frame contains auxiliary information, we can compute an auxiliary vector total over U_F, $\sum_{U_F} \mathbf{x}_k^*$, and use it for calibration. By assumption, the sampled elements belonging to U_P are readily identified. The estimation of $Y_{U_P} = \sum_{U_P} y_k$ can then follow the principles for domain estimation established in Section 6.5. We work with the domain-specific variable y_P such that

$$y_{Pk} = \begin{cases} y_k & \text{if } k \in U_P = U \cap U_F, \\ 0 & \text{otherwise.} \end{cases} \tag{14.2}$$

In the calibration weighting approach, the estimator of Y_{U_P} becomes

$$\hat{Y}_{U_P W} = \sum_{r_F} w_k y_{Pk} = \sum_{r_P} w_k y_k \tag{14.3}$$

where

$$w_k = d_k v_k, \qquad v_k = 1 + \left(\sum_{U_F} \mathbf{x}_k^* - \sum_{r_F} d_k \mathbf{x}_k^* \right)' \left(\sum_{r_F} d_k \mathbf{z}_k^* (\mathbf{x}_k^*)' \right)^{-1} \mathbf{z}_k^* \tag{14.4}$$

for $k \in r_F = r_P \cup r_{\backslash P}$. The standard specification is $\mathbf{z}_k^* = \mathbf{x}_k^*$. The estimator (14.3) is a weighted sum of the y_k-values observed for the responding persisters; y_k-values for nonpersisters, if any are observed, do not count. But $\mathbf{x}_k^*$-values for nonpersisters are allowed to influence the weights. The following example shows the form of $\hat{Y}_{U_P}$W for a simple situation.

Example 14.1.   A commonly used estimator of the persister total

In many common survey designs, the frame population U_F is divided into strata, U_{Fh}, $h = 1, \dots, H$, and a sample s_{Fh} is drawn from U_{Fh} by SI. (Whenever necessary in order to identify a stratum, we add the index h to the notation specified in Section 14.1.) The design weights are then $d_k = N_{Fh}/n_{Fh}$ for $k \in U_{Fh}$. Let $m_{Fh} = m_{Ph} + m_{\backslash Ph}$ be the number of respondents in stratum h, where m_{Ph} and $m_{\backslash Ph}$ are the respective sizes of r_{Ph} and $r_{\backslash Ph}$. A commonly used estimator of Y_{U_P} is

$$\hat{Y}_{U_P}\text{W} = \sum_{h=1}^{H} \frac{N_{Fh}}{m_{Ph} + m_{\backslash Ph}} \sum_{r_{Ph}} y_k = \sum_{h=1}^{H} \frac{N_{Fh}}{m_{Fh}} \sum_{r_{Fh}} y_{Pk}. \qquad (14.5)$$

It should be emphasized that (14.5) sets the limited objective of estimating the persister set total Y_{U_P}, rather than the whole target population total. Although $\hat{Y}_{U_P}$W is in general biased for Y_{U_P}, it is nearly unbiased if the response probability is constant in every stratum, because $(N_{Fh}/m_{Fh}) \sum_{r_{Fh}} y_{Pk}$ is then nearly unbiased for $\sum_{U_{Fh}} y_{Pk}$. If $\hat{Y}_{U_P}$W were viewed as an estimator for the whole target population total $Y_U = \sum_U y_k$, it would give an underestimate – all the more so if the undercoverage is considerable. This would happen even though the response probability may be constant in every stratum. A compensation term, as discussed in step (1b), must be added.

The estimator (14.5) is obtainable as a special case of the general estimator (14.3). We get it by taking the auxiliary vector to be the stratum identifier $\mathbf{x}_k^* = (\gamma_{1k}, \dots, \gamma_{hk}, \dots, \gamma_{Hk})' = \mathbf{z}_k^*$, where, for $h = 1, \dots, H$, $\gamma_{hk} = 1$ for $k \in U_{Fh}$ and $\gamma_{hk} = 0$ otherwise. The required information input is $\sum_{U_F} \mathbf{x}_k^* = (N_{F1}, \dots, N_{Fh}, \dots, N_{FH})'$, the vector of sizes of the frame strata. The matrix to be inverted in (14.4) is diagonal, so we easily find

$$v_k = \frac{n_{Fh}}{m_{Ph} + m_{\backslash Ph}}$$

for $k \in r_{Fh} = r_{Ph} \cup r_{\backslash Ph}$ and therefore

$$w_k = \frac{N_{Fh}}{n_{Fh}} \frac{n_{Fh}}{m_{Ph} + m_{\backslash Ph}} = \frac{N_{Fh}}{m_{Ph} + m_{\backslash Ph}}.$$

We have thus obtained the weights of the estimator (14.5).   □

We turn to variance estimation for $\hat{Y}_{U_P}$W given by (14.3) and (14.4). We can use Proposition 11.1 with $\mathbf{x}_k = \mathbf{x}_k^*$ and r replaced by r_F. Recall that the variance estimator technique in Proposition 11.1 was developed with the aid of proxies for the unknown influences $\phi_k = 1/\theta_k$. In the present context with coverage errors, we must work with different proxies than those in Proposition 11.1. Here we use instead $\hat{\phi}_k = v_k$ given by (14.4). The resulting variance estimator is

$$\hat{V}(\hat{Y}_{U_P\mathrm{W}}) = \hat{V}_{\mathrm{SAM}} + \hat{V}_{\mathrm{NR}} \tag{14.6}$$

where

$$\hat{V}_{\mathrm{SAM}} = \sum\sum_{r_F} (d_k d_\ell - d_{k\ell})(v_k \hat{e}_{Pk})(v_\ell \hat{e}_{P\ell}) - \sum_{r_F} d_k(d_k - 1)v_k(v_k - 1)(\hat{e}_{Pk})^2 \tag{14.7}$$

and

$$\hat{V}_{\mathrm{NR}} = \sum_{r_F} d_k^2 v_k(v_k - 1)\hat{e}_{Pk}^2 \tag{14.8}$$

where v_k is given by (14.4), and

$$\hat{e}_{Pk} = y_{Pk} - (\mathbf{x}_k^*)'\mathbf{B}_{(P)r_F;dv}^* \tag{14.9}$$

with

$$\mathbf{B}_{(P)r_F;dv}^* = \left(\sum_{r_F} d_k v_k \mathbf{z}_k^*(\mathbf{x}_k^*)'\right)^{-1}\sum_{r_F} d_k v_k \mathbf{z}_k^* y_{Pk}. \tag{14.10}$$

We note that for a nonpersister, the residual has the form $\hat{e}_{Pk} = -(\mathbf{x}_k^*)'\mathbf{B}_{(P)r_F;dv}^*$.

14.3.   DIRECT ESTIMATION OF THE TARGET POPULATION TOTAL

When the auxiliary vector $\mathbf{x}_k = \mathbf{x}_k^*$ is specified in the frame for all elements $k \in U_F$, the frame auxiliary total $\sum_{U_F} \mathbf{x}_k^*$ is easily derived by summation. But the aim is to estimate the target population total $Y_U = \sum_U y_k$, not $\sum_{U_F} y_k$. The weights ought to be calibrated to $\sum_U \mathbf{x}_k^*$ instead. In the presence of coverage errors, the two vector totals do not in general agree.

We need to approximate $\sum_U \mathbf{x}_k^*$. Let $\tilde{\mathbf{X}}$ denote an approximation. If the effect of overcoverage can be assumed to roughly cancel the effect of undercoverage, it is realistic to take $\tilde{\mathbf{X}} = \sum_{U_F} \mathbf{x}_k^*$. Otherwise, one may be hard pressed to quantify a close approximation to $\sum_U \mathbf{x}_k^*$. At the estimation stage, when calibrated weights are to be produced, it is nevertheless critically important to have a close approximation. If $\tilde{\mathbf{X}}$ is considerably in error, the quality of the calibrated weights may be jeopardized. The resulting estimates may be considerably biased.

If the frame U_F represents a frequently updated register, the total $\tilde{\mathbf{X}} = \sum_{U_F} \mathbf{x}_k^*$, obtained by 'most recent count' in the register, will always be a rather good approximation to the desired total $\sum_U \mathbf{x}_k^*$. This is the case, for example, with the total population registers in the Scandinavian countries. The situation is less satisfactory in many other circumstances. A business register, for example, may be updated to give a near-complete coverage only once or twice during a calendar year, so $\sum_{U_F} \mathbf{x}_k^*$ is only on rare occasions a satisfactory approximation to $\sum_U \mathbf{x}_k^*$. At other times, assumptions would have to be invoked, so as to compensate for a perceived difference between $\sum_{U_F} \mathbf{x}_k^*$ and $\sum_U \mathbf{x}_k^*$.

The calibration approach leads to estimating $Y_U = \sum_U y_k$ by

$$\hat{Y}_{UW} = \sum_{r_P} w_k y_k \tag{14.11}$$

where

$$w_k = d_k v_k, \qquad v_k = 1 + \left(\tilde{\mathbf{X}} - \sum_{r_P} d_k \mathbf{x}_k^* \right)' \left(\sum_{r_P} d_k \mathbf{z}_k^* (\mathbf{x}_k^*)' \right)^{-1} \mathbf{z}_k^*. \tag{14.12}$$

The estimator is a weighted sum of y_k-data observed for the responding persisters. Individual values $\mathbf{x}_k^*$ are also required for the responding persisters. The weights w_k have the calibration property $\sum_{r_P} w_k \mathbf{x}_k^* = \tilde{\mathbf{X}}$. Provided that $\hat{\mathbf{X}}$ is a close approximation, the resulting estimator $\hat{Y}_{UW}$ may effectively control all three types of error, sampling error, nonresponse error and coverage error. This presupposes the existence of a powerful auxiliary vector, $\mathbf{x}_k^*$.

Example 14.2.   A commonly used estimator of the target population total

Assume that the sample s_F is drawn from the frame U_F by STSI, as described in Example 14.1. It makes sense to construct an estimator using the known stratum counts N_{Fh}. An easily understood and commonly used estimator of Y_U is

$$\hat{Y}_{UW} = \sum_{h=1}^{H} \frac{N_{Fh}}{m_{Ph}} \sum_{r_{Ph}} y_k. \tag{14.13}$$

This is the special case obtained from the general formulas (14.11) and (14.12) when the auxiliary vector is the stratum identifier vector $\mathbf{x}_k^* = (\gamma_{1k}, \ldots, \gamma_{hk}, \ldots, \gamma_{Hk})' = \mathbf{z}_k^*$. But two requirements are needed to claim unbiasedness for $\hat{Y}_{UW}$. The first is that the population stratum sizes are unchanged between the time of sampling (the reference time for U_F) and the time of estimation (the reference time for U). In each stratum, the overcoverage is assumed to compensate exactly for the undercoverage. If this holds, we can compute reliable calibrated weights using $\tilde{\mathbf{X}} = (N_{F1}, \ldots, N_{Fh}, \ldots, N_{FH})' = \sum_{U_F} \mathbf{x}_k^*$. The second requirement is that all elements in stratum h, persisters as well as nonpersisters, respond with the same probability.

A little algebra shows that v_k, given by (14.12), takes the form $v_k = n_{Fh}/m_{Ph}$ for all $k \in r_{Ph}$, so that the weight of element k becomes $w_k = (N_{Fh}/n_{Fh})(n_{Fh}/m_{Ph}) = N_{Fh}/m_{Ph}$ for all $k \in r_{Ph}$. This is precisely the weight of y_k in (14.13).

□

We consider the variance estimation for $\hat{Y}_{UW}$ given by (14.11). Proposition 11.1 is useful again, with certain modifications. The variance estimator (11.3) was derived with the aid of proxies for the influences $\phi_k = 1/\theta_k$. Here we indicate two alternatives for the proxies, denoted in general by $\hat{\phi}_k$.

From Proposition 11.1.1, with r replaced by r_P, we obtain

$$\hat{V}(\hat{Y}_{UW}) = \hat{V}_{SAM} + \hat{V}_{NR} \tag{14.14}$$

where

$$\hat{V}_{SAM} = \sum\sum\nolimits_{r_P} (d_k d_\ell - d_{k\ell})(\hat{\phi}_k \hat{e}_k)(\hat{\phi}_\ell \hat{e}_\ell) - \sum\nolimits_{r_P} d_k(d_k - 1)\hat{\phi}_k(\hat{\phi}_k - 1)\hat{e}_k^2$$
(14.15)

and

$$\hat{V}_{NR} = \sum\nolimits_{r_P} d_k^2 \hat{\phi}_k(\hat{\phi}_k - 1)\hat{e}_k^2$$
(14.16)

where

$$e_k = y_k - (\mathbf{x}_k^*)'\mathbf{B}_{r_P;d\hat{\phi}}^*$$
(14.17)

with

$$\mathbf{B}_{r_P;d\hat{\phi}}^* = \left(\sum\nolimits_{r_P} d_k \hat{\phi}_k \mathbf{z}_k^*(\mathbf{x}_k^*)'\right)^{-1}\sum\nolimits_{r_P} d_k \hat{\phi}_k \mathbf{z}_k^* y_k.$$
(14.18)

We suggest two alternatives for the proxies $\hat{\phi}_k$ in (14.14)–(14.18). The choice between them is contingent on the availability of certain information.

1   Use $\hat{\phi}_k = v_k$ given by (14.4). An advantage with this procedure is that it dodges the difficulty with a set of nonresponding persisters, o_P, that perhaps cannot be identified. We do not need to know o_P to compute the proxies v_k.

A potential weakness with $\hat{\phi}_k = v_k$ in alternative 1 is that it uses elements in the overcoverage. These elements may experience some or all items on the question-naire as irrelevant, since they are not targeted for the survey. Consequently, they have a greater probability of becoming nonrespondents. Therefore it is question-able to involve the set of elements $r_{\setminus P}$ in the computation of proxies.

2   In (14.15)–(14.18), use $\hat{\phi}_k = v_{Pk}$, where

$$v_{Pk} = 1 + \left(\sum\nolimits_{r_P \cup o_P} d_k \mathbf{x}_k^* - \sum\nolimits_{r_P} d_k \mathbf{x}_k^*\right)'\left(\sum\nolimits_{r_P} d_k \mathbf{z}_k^*(\mathbf{x}_k^*)'\right)^{-1}\mathbf{z}_k^*$$
(14.19)

for $k \in r_P$. The calibration in (14.19) from the set r_P to the set $r_P \cup o_P$ seems reasonable. A potential practical difficulty is that this alternative requires an identification of the set of nonresponding persisters o_P.

14.4.   A CASE STUDY

This section summarizes the steps taken to redesign the estimation method in Statistics Sweden's survey on 'Transition from upper secondary school to higher education'. We call it the School Survey. Its target population consists of the students in the final year of the three-year Swedish secondary school system. The redesign resulted in a change from a 'traditional estimator' to estimation by calibration, using (14.11). We comment on methodological issues arising in the process and on the effects on the survey estimates of the changeover. The example points out some problems that are fairly typical of a survey with coverage errors.

The School Survey has many study variables (y-variables). Important ones are: (a) the intentions of third-year secondary school students with regard to pursuing studies at university; (b) the university programmes viewed by these students as the most attractive and interesting. Since the survey is conducted yearly, an important objective is also to monitor the evolution over time of students' attitudes and preferences with regard to university education.

The School Survey is carried out yearly among third-year students. The school year lasts from August of year t until May in the year $t + 1$. For the survey with reference year $t + 1$ (called the $t + 1$ survey), a stratified sample of students is selected in September of year t. Data are collected by mail from the selected students during October and November.

An administrative register covering all third-year students becomes available in the month of January of year $t + 1$. Among the register variables are gender, region of the school, and the study programme in which a student is enrolled. Some of the study variables are also register variables. The values of the latter variables will thus be available at a later date for all students in the target population. This register is finalized in April of year $t + 1$, but it is essentially already complete in January. This is much too late for the September year t selection of the sample for the $t + 1$ survey. Consequently, the September sample must be drawn from the frame (the register) that lists the second-year students in year t.

A considerable number of students abandon their studies at the end of the second year. They represent frame overcoverage. Students added to the target population (usually due to re-entry into the school system after temporary dropout) represent frame undercoverage. In a typical year, the overcoverage is considerably greater than the undercoverage. It is clear from the outset that the assumption of an undercoverage that roughly compensates for the overcoverage does not hold. Estimation methods that rely on that assumption are likely not to work well.

The School Survey results are subject to sampling error, nonresponse error and frame deficiency error. The last two errors are particularly troublesome. The survey nonresponse is high, at about 25 %. An analysis shows clearly that nonresponse does not occur at random. It is related to several of the important study variables and to important register variables. The nonresponse is thus likely to severely bias the survey results, unless effective weighting can be developed to counteract the tendency to bias.

Under the old survey design, the following estimator was used:

$$\hat{Y}_U = \sum_{h=1}^{H} \frac{\hat{N}_h}{m_{Ph}} \sum_{r_{Ph}} y_k \qquad (14.20)$$

where

$$\hat{N}_h = N_{Fh} \frac{m_{Ph}}{m_{Ph} + m_{\backslash Ph}}. \qquad (14.21)$$

The notation is as in Examples 14.1 and 14.2. There was no clear indication as to why this estimator had come into use at some time in the past. It was viewed as 'traditional'. One notes that (14.20) can be derived from (14.11) and (14.12) if one accepts $\hat{N}_h$ in (14.21) as a reasonable approximation to the size N_h of stratum h of the target population, and $\check{\mathbf{X}}$ defined as the H-vector composed of the quantities $\hat{N}_h$, $h = 1, \ldots, H$.

However, the survey manager had noted clear signs that (14.20) overestimates at least some of the totals. This could be seen by comparing the survey results with the register. The survey estimates computed by (14.20) become available for the study variables after a completed data collection in March $t + 1$. 'True values' of the study variables that are also register variables become available from the finalized register in April $t + 1$. A comparison of the estimates obtained by (14.20) with the 'truth' had shown clear signs of overestimation.

The response propensity is very low among nonpersisters, the students who drop out. Thus, $m_{\backslash Ph}$ is very small, and the estimator (14.20) yields numbers that are very close to what the estimator (14.13) would produce. This latter estimator is 'at a correct level' if $N = N_F$. But $N = N_F$ is far from holding true in this survey. The frame size N_F is known to be considerably greater than the target population size N. It is not surprising, then, that past experience with (14.20) has pointed to a clear and unacceptable overestimation.

Another comparison is of interest: (14.20) is identical to (14.5), described in Section 14.2 as an unbiased estimator of the persister total, that is, the total of the domain $U \cap U_F$, provided all elements in stratum h, persisters as well as nonpersisters, respond with the same probability. When that condition holds, (14.5) estimates the persister total well, but underestimates the total of the target population U. But instead of an anticipated underestimation, (14.5), or equivalently (14.20), overestimates the targeted totals. There are two reasons: (i) rather than being constant within strata for all elements, the response probability is much lower among nonpersisters; and (ii) N is considerably smaller than N_F.

For these reasons, the decision was taken for the $t + 1 = 2003$ survey to change over to a carefully planned calibration estimator of the form (14.11). The prime objective thus became to identify a suitable vector $\mathbf{x}_k^*$ and a suitable information input $\tilde{\mathbf{X}}$ for the weight computation defined by (14.12). The rest of the example describes the identification of the auxiliary variables to be included in $\mathbf{x}_k^*$. We also comment on the improvements in the survey results resulting from the change of estimator.

The frame population for the 2003 survey was stratified by region, second-year study programme and gender. Some collapsing of study programmes proved necessary to avoid extremely small strata. The determination of strata and the allocation of sample to strata were guided by an objective of obtaining precise estimates in especially important domains of interest. A sample of 9023 students was selected by STSI in September 2002.

The computation of calibrated weights for the estimation has to be completed at the latest in the second week of January in order to guarantee a timely production and publication of survey results. The auxiliary vector and the (approximate) auxiliary total vector $\tilde{\mathbf{X}}$ in (14.12) must therefore be decided on by that time.

The register is finalized in April, but an almost identical preliminary version is available in January. In the comments that follow we compare the results of the 2003 School Survey with those obtained from the register as it exists in January. It can be taken as the 'truth' since it differs only marginally from the final April register.

A student is identified in the register by his or her personal identification number. This allowed the redesign team to match the register of second-year students with the January version of the third-year register and to identify the overcoverage set, the undercoverage set and the sets r_p and o_p. It thus became possible to compute the weights v_{Pk} given by (14.19). These were used in the computation of the variance estimator given by (14.14)–(14.18). This variance estimation gave satisfactory results.

The search for variables to include in the auxiliary vector $\mathbf{x}_k^*$ was guided by the principles in Section 10.2. (The indicators in Section 10.6 were not used at the time.) The January third-year register contains several potential auxiliary variables. Important domains are those defined by gender, region and study programme. Therefore these three categorical variables strongly suggest themselves for inclusion in the auxiliary vector, in line with Principle 3. In addition, country of birth was included, because it turned out to be an important auxiliary variable from the perspective of Principles 1 and 2.

The redesign team noted that two other important administrative sources could be brought into the picture. By personal identification number matching it was possible to import from another register the variable 'final mark', which is the student's summary mark, combining all subjects of study, received at the end of grade 9, prior to entering the final three years of school. It turned out to be a powerful auxiliary variable. The other important administrative source was the TPR, as described in Example 2.1. From this source the team could extract several powerful variables concerning the identified parent (the more senior one, if two) of a sampled student. These parental variables were level of education, income and civil status.

The final choice of auxiliary variables was prepared through an analysis similar to the one in Example 3.5. Details are omitted here. Table 14.1 shows some survey results concerning the final mark auxiliary variable, presented as a grouped variable, based on intervals. In view of Principle 1 we are interested in the relation of final mark to response propensity. Table 14.1 indicates a strong relationship. In view of Principle 2 we are interested in how final mark relates to important study variables. Table 14.1 shows this relation to be strongly pronounced for the important study variable of intention to pursue university studies. Final mark is thus an important auxiliary variable from the perspective of both Principle 1 and Principle 2.

The final decision was to let the auxiliary vector consist of seven categorical auxiliary variables. For weight computation, they were strung out "side by side" in the manner of (7.19). This implies a calibration on the marginal frequencies

Table 14.1 Proportion of respondents and proportion of students intending to pursue university studies, both as a function of final mark. Data from Statistics Sweden's 2003 School Survey.

Final mark in grade 9	0–160	161–200	201–240	241–320
Response rate in %	63.8	68.1	76.9	84.6
Proportion intending to pursue university studies	23.7	27.9	55.9	78.2

of each categorical variable. The vector $\tilde{\mathbf{X}}$ consists of the marginal counts taken from the January register.

What main conclusions emerge from changing the estimation in the survey from the old method (14.20) to the new method? Totals as well as proportions are estimated in the survey. The estimates of totals undergo considerable changes. It seems safe to assume that these changes go in the direction of reduced estimation error. For the totals of the register variables, exact estimates are obtained because of the calibration property.

The estimates of proportions undergo little change as a result of the change in estimation method. One plausible explanation is that the change of estimator causes similar moves in the numerator and in the numerator, with little change in the ratio as a result. The estimated variances for proportions were not much reduced by the change to the new estimation technique. This is not entirely unexpected, considering that the stratified sampling design already accounts for some powerful auxiliary information.

Without question, the estimation in this survey was improved by the transition to calibration estimation in line with (14.11) and (14.12). Users had spotted weaknesses in the old method: (a) nonresponse analyses had suggested that the old method gave biased estimates; and (b) the coverage errors were known to be substantial and influential. It is safe to conclude that the new estimation method reduces the estimation error, especially for the estimation of totals. Another positive outcome was improved user trust in the estimates.

References

Andersson, C. and Nordberg, L. (1998) *CLAN97 – A SAS-program for Computation of Point- and Standard Error Estimates in Sample Surveys.* Örebro: Statistics Sweden.

ANGOSS Software (1995) *KnowledgeSEEKER IV for Windows – User's Guide.* ANGOSS Software International Limited.

Atmer, J., Thulin, G. and Bäcklund, S. (1975) Coordination of samples with the JALES technique. *Statistisk Tidskrift*, **13**, 343–350.

Bankier, M. D., Rathwell, S. and Majkowski, M. (1992) Two step generalized least squares estimation in the 1991 Canadian Census. *Proceedings of Survey Research Methods Section*, American Statistical Association, pp. 764–769

Bethlehem, J. G. (1988) Reduction of nonresponse bias through regression estimation. *Journal of Official Statistics*, **4**, 251–260.

Bethlehem, J. (1999) Cross-sectional research. In H. J. Adèr and G. J. Mellenbergh (eds.), *Research Methodology in the Social, Behavioural and Life Sciences*, pp. 110–142. London: Sage.

Bethlehem, J. G. and Kersten, H. M. P. (1985) On the treatment of nonresponse in sample surveys. *Journal of Official Statistics*, **1**, 287–300.

Bethlehem, J. and Schouten, B. (2004) Nonresponse adjustment in household surveys. Discussion paper 04007. Voorburg: Statistics Netherlands.

Breidt, F. J. and Opsomer, J. D. (2000) Local polynomial regression estimators in survey sampling. *Annals of Statistics*, **28**, 1026–1053.

Brewer, K. R. W., Early, L. J. and Joyce, S. F. (1972) Selecting several samples from a single population. *Australian Journal of Statistics*, **14**, 231–239.

Caron, N. (1998) Le logiciel POULPE: aspects méthodologiques. *Actes des Journées de Méthodologie*, INSEE, Paris.

Caron, N., Deville, J. C. and Sautory, O. (1998) Estimation de précision de données issues d'enquêtes: document méthodologique sur le logiciel POULPE. Document de travail de la Direction des Statistiques Démographiques et Sociales No. 9806, INSEE, Paris.

Cochran, W. G. (1977) *Sampling Techniques*, 3rd edn. New York: Wiley.

de Leeuw, E. and de Heer, W. (2002) Trends in household survey nonresponse: A longitudinal and international comparison. In R. M. Groves, D. A. Dillman, J. L. Eltinge and J. A. R. Little (eds), *Survey Nonresponse*. New York: Wiley.

Deville, J. C. (1993) Une formule universelle d'estimation de variance. Internal report. INSEE, Paris.

Deville, J. C. (2000) Generalized calibration and application to weighting for non-response. In J. G. Bethlehem and P. G. M. van der Heijden (eds), *COMPSTAT – Proceedings in Computational Statistics*, pp. 65–76. Heidelberg: Physica-Verlag.

Deville, J. C. and Särndal, C. E. (1992) Calibration estimators in survey sampling. *Journal of the American Statistical Association*, **87**, 376–382.

Deville, J. C., Särndal, C. E. and Sautory, O. (1993) Generalized raking procedures in survey sampling. *Journal of the American Statistical Association*, **88**, 1013–1020.

Djerf, K. (1997) Effects of post-stratification on the estimates of the Finnish Labour Force Survey. *Journal of Official Statistics*, **13**, 29–39.

Djerf, K. (2000) Properties of some estimators under unit nonresponse. Research Report no. 231, Statistics Finland, Helsinki.

Dufour, J., Gagnon, F., Morin, Y., Renaud, M. and Särndal, C. E. (2001) A better understanding of weight transformation through a measure of change. *Survey Methodology*, **27**, 97–108.

Ekholm, A. and Laaksonen, S. (1991) Weighting via response modeling in the Finnish Household Budget Survey. *Journal of Official Statistics*, **3**, 325–337.

Estevao, V. and Särndal, C. E. (2002) The ten cases of auxiliary information for calibration in two-phase sampling. *Journal of Official Statistics*, **18**, 233–255.

Estevao, V. M., Hidiroglou, M. A. and Särndal, C. E. (1995) Methodological principles for a generalized estimation system at Statistics Canada. *Journal of Official Statistics*, **11**, 181–204.

Folsom, R. E. (1991) Exponential and logistic weight adjustments for sampling and nonresponse error reduction. *Proceedings of the Social Statistics Section*, American Statistical Association, pp. 197–202.

Fuller, W. A. (2002) Regression estimation for survey samples. *Survey Methodology*, **28**, 5–23.

Fuller, W. A., Loughin, M. M. and Baker, H. D. (1994) Regression weighting in the presence of nonresponse with application to the 1987–1988 nationwide Food Consumption Survey. *Survey Methodology*, **20**, 75–85.

Gabler, S. and Häder, S. (1999) Representive weights and imputation for the 1997 German ISSP: An application of the conditional minimax principle. Paper presented at the International Conference on Survey Nonresponse, Portland, OR.

Groves, R. M. and Couper, M. P. (1998) *Nonresponse in Household Interview Surveys*. New York: Wiley-Interscience.

Hàjek, J. (1981) *Sampling from a Finite Population*. New York: Dekker.

Holt, D. and Elliot, D. (1991) Methods of weighting for unit non-response. *The Statistician*, **40**, 333–342.

Hörngren, J. (1992) The use of registers as auxiliary information in the Swedish Labour Force Survey. R&D Report no. 1992:13, Statistics Sweden, Stockholm.

Kalton, G. and Kasprzyk, D. (1986) The treatment of missing data. *Survey Methodology*, **12**, 1–16.

Kalton, G. and Maligalig, D. S. (1991) A comparison of weighting adjustment for nonresponse. *Proceedings of the Bureau of the Census Annual Research Conference*, pp. 409–428.

Kass, G. V. (1980) An exploratory technique for investigating large quantities of categorical data. *Applied Statistics*, **29**, 119–127.

Kish, L. (1979) Samples and censuses. *International Statistical Review*, **47**, 99–110.

Kish, L. and Anderson, D. W. (1978) Multivariate and multipurpose stratification. *Journal of the American Statistical Association*, **73**, 24–34.

Kott, P. S. (1994) A note on handling nonresponse in sample surveys. *Journal of the American Statistical Association*, **89**, 693–696.

Lee, H., Rancourt, E. and Särndal, C. E. (2001) Variance estimation from survey data under single imputation. In R. M. Groves, D. A. Dillman, J. L. Eltinge and J. A. R. Little (eds), *Survey Nonresponse*. New York: Wiley.

Lehtonen, R. and Veijanen, A. (1998) Logistic generalized regression estimators. *Survey Methodology*, **24**, 51–55.

Lindström, H. (1983) *Non-response Errors in Sample Surveys*, Urval 16. Stockholm: Statistics Sweden.

Lindström, H. and Lundström, S. (1974) A method to discuss the magnitude of the non-response error. *Statistisk Tidskrift*, **12**, 505–520.

Little, R. J. A. (1986) Survey nonresponse adjustments for estimates of means. *International Statistical Review*, **54**, 139–157.

Little, R. J. A. (1988) Missing-data adjustment in large surveys. *Journal of Business & Economic Statistics*, **6**, 287–296.

Little, R. J. A. and Rubin, D. B. (1987) *Statistical Analysis with Missing Data*. New York: Wiley.

Lohr, S. L. (1999) *Sampling: Design and Analysis*. Pacific Grove, CA: Duxbury Press.

Lundström, S. (1996) Kalibrering av vikter i kortperiodisk sysselsättningsstatistik för privat sektor. *Proceedings of the 20th Conference of Nordic Statisticians*, Copenhagen, Denmark, pp. 442–451.

Lundström, S. (1997) Calibration as a standard method for treatment of nonresponse. Ph.D. thesis, Stockholm University.

Lundström, S. and Särndal, C. E. (1999) Calibration as a standard method for treatment of nonresponse. *Journal of Official Statistics*, **15**, 305–327.

Lundström, S. and Särndal, C. E. (2001) *Estimation in the Presence of Nonresponse and Frame Imperfections*. Örebro: Statistics Sweden.

Montanari, G. E. and Ranalli, M. G. (2003) Nonparametric methods in survey sampling. In M. Vinci, P. Monari, S. Mignani and A. Montanari (eds), *New Developments in Classification and Data Analysis*. Berlin: Springer.

Nascimento Silva, P. L. D. and Skinner, C. J. (1997) Variable selection for regression estimation in finite populations. *Survey Methodology*, **23**, 23–32.

Nieuwenbrook, N. J. and Boonstra, H. J. (2002) Bascula 4.0 for weighting sample survey data with estimation of variances. *Survey Statistician*, Software reviews, July.

Oh, H. L. and Scheuren, F. J. (1983) Weighting adjustment for unit nonresponse. In W. G. Madow, I. Olkin and D. B. Rubin (eds), *Incomplete Data in Sample Surveys*, Vol. 2. New York: Academic Press.

Platek, R. and Gray, G. B. (1983) Imputation methodology. In W. G. Madow, I. Olkin and D. B. Rubin (eds), *Incomplete Data in Sample Surveys*, Vol. 2. New York: Academic Press.

Platek, R., Singh, M. P. and Tremblay, V. (1978) Adjustments for nonresponse in surveys. In N. K. Namboodiri (ed.), *Survey Sampling and Measurement*. New York: Academic Press.

Rizzo, L., Kalton, G. and Brick, J. M. (1996) A comparison of some weighting adjustment methods for panel nonresponse. *Survey Methodology*, **22**, 43–53.

Rosén, B. (1997a) Asymptotic theory for order sampling. *Journal of Statistical Planning and Inference*, **62**, 135–158

Rosén, B. (1997b) On sampling with probability proportional to size. *Journal of Statistical Planning and Inference*, **62**, 159–191

Rubin, D. B. (1978) Multiple imputation in sample surveys – a phenomenological Bayesian approach to nonresponse. *Proceedings of the Section on Survey Research Methods*, American Statistical Association, pp. 20–34

Rubin, D. B. (1983) Conceptual issues in the presence of nonresponse. In W. G. Madow, I. Olkin and D. B. Rubin (eds), *Incomplete Data in Sample Surveys*, Vol. 2. New York: Academic Press.

Rubin, D. B. (1987) *Multiple Imputation for Nonresponse in Surveys*. New York: Wiley.

Rubin, D. B. (1992) Discussion. *Proceedings of the Bureau of the Census Annual Research Conference*, pp. 540–544.

Särndal, C. E. and Swensson, B. (1987) A general view of estimation for two phases of selection with applications to two-phase sampling and nonresponse. *International Statistical Review*, **55**, 279–294.

Särndal, C. E., Swensson, B. and Wretman, J. H. (1992) *Model Assisted Survey Sampling*. New York: Springer-Verlag.

Statistics Sweden (1980) *Räkna med bortfall* (Computation with nonresponse). Stockholm: Statistics Sweden.

Thomsen, I. (1973) A note on the efficiency of weighting subclass means to reduce the effects of non-response when analyzing survey data. *Statistisk Tidskrift*, **11**, 278–285.

Thomsen, I. (1978) A second note on the efficiency of weighting subclass means to reduce the effects of non-response when analyzing survey data. *Statistisk Tidskrift*, **16**, 191–196.

Willimack, D. K., Nichols, E. and Sudman, S. (2002) Understanding unit and item nonresponse in business surveys. In R. M. Groves, D. A. Dillman, J. L. Eltinge and J. A. R. Little (eds), *Survey Nonresponse*, New York: Wiley.

Wright, R. L. (1983) Finite population sampling with multivariate auxiliary information. *Journal of the American Statistical Association*, **78**, 879–884.

Index

WILEY SERIES IN SURVEY METHODOLOGY
Established in Part by WALTER A. SHEWHART AND SAMUEL S. WILKS

Editors: *Robert M. Groves, Graham Kalton, J. N. K. Rao, Norbert Schwarz, Christopher Skinner*

Wiley Series in Survey Methodology covers topics of current research and practical interests in survey methodology and sampling. While the emphasis is on application, theoretical discussion is encouraged when it supports a broader understanding of the subject matter.

The authors are leading academics and researchers in survey methodology and sampling. The readership includes professionals in, and students of, the fields of applied statistics, biostatistics, public policy, and government and corporate enterprises.

BIEMER, GROVES, LYBERG, MATHIOWETZ, and SUDMAN (editors) Measurement Errors in Surveys
CHAMBERS and SKINNER (editors). Analysis of Survey Data
COCHRAN Sampling Techniques, *Third Edition*
COUPER, BAKER, BETHLEHEM, CLARK, MARTIN, NICHOLLS, and O'REILLY (editors) Computer Assisted Survey Information Collection
COX, BINDER, CHINNAPPA, CHRISTIANSON, COLLEDGE, and KOTT (editors) Business Survey Methods
*DEMING Sample Design in Business Research
DILLMAN Mail and Telephone Surveys: The Total Design Method, *Second Edition*
DILLMAN Mail and Internet Surveys: The Tailored Design Method
GROVES and COUPER Nonresponse in Household Interview Surveys
GROVES Survey Errors and Survey Costs
GROVES, DILLMAN, ELTINGE, and LITTLE (editors) Survey Nonresponse
GROVES, BIEMER, LYBERG, MASSEY, NICHOLLS, and WAKSBERG Telephone Survey Methodology
*HANSEN, HURWITZ, and MADOW Sample Survey Methods and Theory, Volume I: Methods and Applications
*HANSEN, HURWITZ, and MADOW Sample Survey Methods and Theory, Volume II: Theory
KISH Statistical Design for Research
*KISH Survey Sampling
KORN and GRAUBARD Analysis of Health Surveys
LESSLER and KALSBEEK Nonsampling Error in Surveys
LEVY and LEMESHOW Sampling of Populations: Methods and Applications, *Third Edition*
LYBERG, BIEMER, COLLINS, de LEEUW, DIPPO, SCHWARZ, TREWIN (editors) Survey Measurement and Process Quality
MAYNARD, HOUTKOOP-STEENSTRA, SCHAEFFER, VAN DER ZOUWEN Standardization and Tacit Knowledge: Interaction and Practice in the Survey Interview
SÄRNDAL and LUNDSTRÖM Estimation in Surveys with Nonresponse
SIRKEN, HERRMANN, SCHECHTER, SCHWARZ, TANUR, and TOURANGEAU (editors) Cognition and Survey Research
VALLIANT, DORFMAN, and ROYALL Finite Population Sampling and Inference: A Prediction Approach

*Now available in a lower priced paperback edition in the Wiley Classics Library.